AF560146

OPERATION RESEARCH
MISCELLANEOUS TOPICS

OPERATION RESEARCH
MISCELLANEOUS TOPICS

By
S.C. Sharma

DISCOVERY PUBLISHING HOUSE
NEW DELHI-110002

First Published – 2006

Reprinted – 2025

ISBN: 978-81-8356-103-7

Operation Research:
Miscellaneous Topics

Published by:

DISCOVERY PUBLISHING HOUSE
4383/4B, Ansari Road, Darya Ganj
New Delhi-110 002 (India)
Phone: +91-11-23279245; 23253475; 43596065
Mobile: +91 9811179893 / +91 9871656464
E-mail: discoverybooksindia@gmail.com
orderdphbooks@gmail.com
namitwasan9@gmail.com
web: www.discoverypublishinggroup.com

Printed at:
Infinity Imaging Systems
Delhi

Preface

This book on Operation Research has been specially written to meet the requirements of the M.Sc., M.Com. and M.B.A. students for all Indian Universities.

The subject matter has been discussed in such a simple way that the students will find no difficulty to understand it. The proof of various theorems and examples has been given with minute details. Each chapter of this book contains complete theory and fairly large number of solved examples, sufficient problems have also been selected from various universities examination papers.

In the preparation of this book large number of books and research papers have been consulted. So no authenticity is claimed.

The author express has gratitude to Mr. Wasan and staff of M/s Discovery Publishing House for their whole hearted co-operation in the publication of this book.

The author tried hard to be accurate and upto date in statement and realises the impressibility of completely avoiding error's. Therefore the author will greatly appropriate having his authentic called to any questionable statement.

Author

Contents

1

Dynamic Programming

INTRODUCTION

Often decision-making process involves several decisions to be taken at different times. For example, problems of inventory control, evaluation of investment opportunities, long-term corporate planning, and so on require sequential decision-making. The mathematical technique of optimizing such a sequence of interrelated decisions over a period of time is called *dynamic programming*. It uses the idea of recursion to solve a complex problem, broken into a series of interrelated (sequential) decision stages (also called *sub-problems*) where the outcome of a decision at one stage affects the decision at each of the next stages. The word *dynamic* has been used because time is explicitly taken into consideration.

Dynamic programming (DP) differs from linear programming in two ways:

(i) In DP, there is no set procedure (algorithm) as in LP that can be used to solve all problems. DP is a technique that allows to break up the given problem into a sequence of easier and smaller sub-problems which are then solved in stages.

(ii) Q LP gives one time period (single stage) solution whereas DP considers decision-making over time and solves each sub-problem optimally.

DYNAMIC PROGRAMMING TERMINOLOGY

Regardless of the type or size of a dynamic programming problem, there are certain terms and concepts that are common in every problem.

Stage: The dynamic programming problem can be decomposed or divided into a sequence of smaller sub-problems called *stages*. At each stage there are a number of decision alternatives (courses of action) and a decision is made by selecting the most suitable alternative. Stages very often represent different time periods in the planning period of the problem, places, people or other entities. For example, in the replacement probably, each year is a stage, in the salesman allocation problem, each territory represents a stage.

State: Each stage in a dynamic programming problem has a certain number of states associated with it. These states represent various conditions of the decision process at a stage. The variables which specify the condition of the decision process or describe the status of the system at a particular stage are called *state variables.* These variables provide information for analysing the possible effects that the current decision, could have upon future courses of action. At any stage of the decision-making process there could be a finite or infinite number of states. For example, a specific city is referred to as state variable in any stage of the shortest route problem.

Return function : At each stage, a decision is made which can affect the state of the system at the next stage and help in arriving at the optimal solution at the current stage. Every decision that is made has its own merit in terms of worth or benefit associated with it and can be described in an algebraic equation form. This equation is generally called a return function, since for every set of decisions, a return on each decision is obtained. This *return function* in general depends on the *state variable* as well as the decision made at a particular stage. An *optimal policy* or *decision* at a stage yields optimal (maximum or minimum) return for a given value of the state variable.

Figure 1.1 depicts the decision alternatives known at each stage for their evaluation. The range of such decision alternatives and their associated returns at a particular stage is a function of the state input to the stage itself. The state input to a stage is the output from the previous (larger number) stage and the previous stage output is a function of the state input to itself, and the decision taken at that stage. Thus to evaluate any stage we need to know the values of the state input to it (there may be more than one state inputs to a stage) and the decision alternatives and their associated returns at the stage.

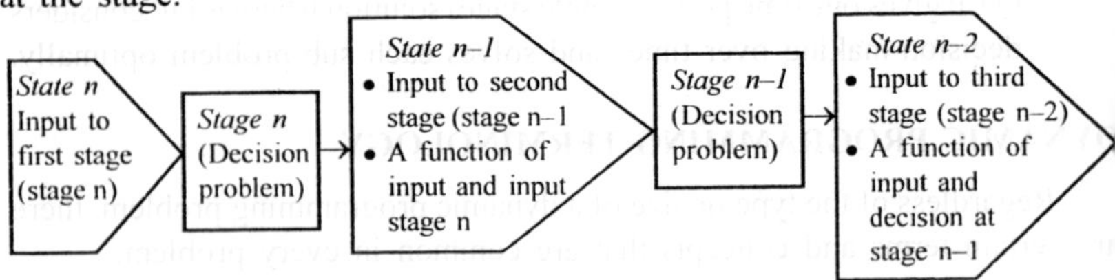

Fig. 1.1 : Information Flow Between Stages

For a multi-stage decision process, functional relationship among state, stage and decision may be described as shown in Fig. 1.2.

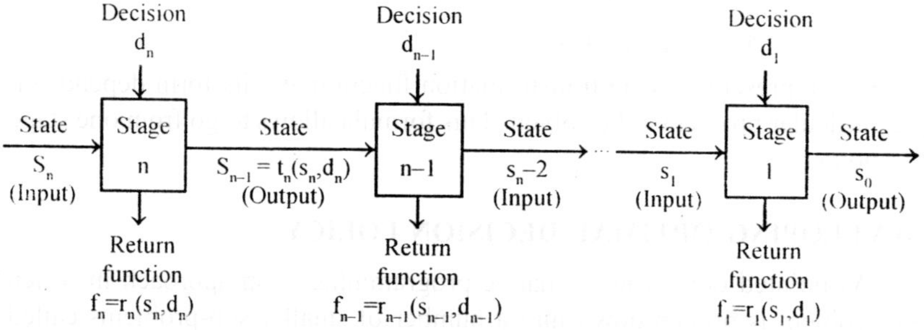

Fig. 1.2 : Functional Relationship Among Components of a DP Model

where n = stage number

s_n = state input to stage n from stage n + 1. Its value is the status of the system resulting from the previous (n + 1) stage decision.

d_n = decision variable at stage n (independent of previous stages). It represents the range of alternatives that can be selected from when making a decision at stage n.

f_n = r_n (s_n, d_n) = return (or objective) function for stage n.

Further suppose that there are n stages at which a decision is to be made. These n stages are all interconnected by the relationship (called transition *function*):

$$s_{n-1} = s_n * d_n.$$

that is,(Ouput at stage n) – (Input to state n) * (Decision at stage n).

where * represents any mathematical operation, namely addition, subtraction, division or multiplication. The units of s_n, d_n and s_{n-1} must be homogeneous.

It can be seen that at each stage of the problem, there are two inputs: state (variable) s_n and decision (variable) d_n. The state (variable) is the state input which relates the present stage back to the previous stage. For example, the current state s_n, provides complete information about various possible conditions in which the problem is to be solved when there are n stages to go. The decision d_n is made at stage n for optimizing the total return over the remaining n–1 stages. The decision d_n which optimizes the output at stage n produces two outputs: (i) the return function r_n (s_n, d_n) and (ii) the new state variable s_{n-1}.

The return function which is expressed as function of the state variable, s_n and the decision (variable), d_n indicates about the state of the process at the beginning of the next stage (stage n–1), and is denoted by *transition function* (state transformation)

$$s_{n-1} = t_n (s_n, d_n),$$

where t_n represents a state transformation function and its form depends on the particular problem to be solved. This formula allows to go from one stage to another.

DEVELOPING OPTIMAL DECISION POLICY

As pointed out earlier, dynamic programming is an approach in which the problem is broken down into a number of smaller sub-problems called *stages*. These sub-problems are then solved sequentially until the original problem is finally solved. A particular sequence of alternatives (courses of action) adopted by the decision-maker in a multi-stage decision problem is called a policy. The optimal policy, therefore, is the sequence of alternatives that achieves the decision-maker's objective. The solution of a dynamic programming problem is based upon *Bellman's principle of optimality* (recursive optimization technique) which states:

The optimal policy must be one such that, regardless of how a particular state is reached, all later decisions (choices) proceeding from that state must be optimal.

Based on this principle of optimality, we find the best policy by solving one stage at a time, and then sequentially adding a series of one-stage-problems that are solved until the overall optimum of the initial problem is obtained. The solution procedure is based on a *backward induction process and forward induction process*. In the first process, the problem is solved by solving the problem in the last stage and working backwards towards the first stage, making optimal decisions at each stage of the problem. In certain cases, the second process is used to solve a problem by first solving the initial stage of the problem and working towards the last stage, making an optimal decision at each stage of the problem.

It was also mentioned earlier that the exact recursion relationship would vary according to the nature of the problem to be solved by dynamic programming. It is clear from the above discussion that one stage return is given by

$$f_1 = r_1 (s_1, d_1)$$

and the optimal value of f_1 under the state variable s_1 can be obtained by selecting a suitable decision variable d_1. That is,

$$f_1^*(s_1) = \underset{d_1}{\text{Out}} \{r_1(s_1, d_1)\}.$$

The range of d_1 is determined by s_1, but s_1 is determined by what has happened in Stage 2. Then in Stage 2 the return function will take the form:

$$f_1^* (s_2) = \underset{d_1}{\text{Out}} \left\{ r_2 (s_2) * f_1^* (s_1) \right\}; \; s_1 = t_2 (s_2, d_2)$$

By continuing the above logic recursively for a general n stage problem, we have

$$f_2^* (s_n) = \underset{d_n}{\text{Out}} \left\{ r_n (s_n, d_n) * f_2^* (s_{n-1}) \right\}; \; s_{n-1} = t_n (s_n, d_n).$$

Here the symbol * denotes any mathematical relationship between s_n and d_n, including addition, subtraction, multiplication, etc.

The General Algorithm

The procedure for solving a problem by using the dynamic programming approach can be summarized in the following steps:

Step 1. Identify the problem decision variables and specify objective function to be optimized under certain limitations, if any.

Step 2. Decompose (or divide) the given problem into a number of smaller sub-problems (or stages). Identify the state variables at each stage and write down the transformation function as a function of the state variable and decision variable at the next stage.

Step 3. Write down a general recursive relationship for computing the optimal policy. Decide whether to follow the forward or the backward method to solve the problem.

Step 4. Construct appropriate tables to show the required values of the return function at each stage as shown in Table 1.1.

Step 5. Determine the overall optimal policy or decisions and its value at each stage. There may be more than one such optimal policy.

Table 1.1 : Stage 1

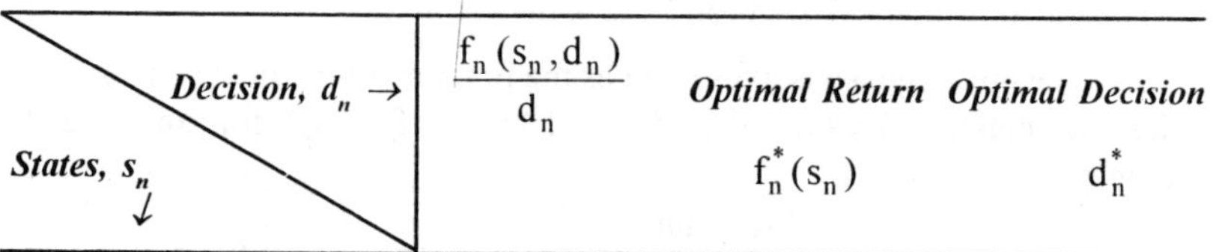

Decision, d_n → / *States,* s_n ↓	$\dfrac{f_n (s_n, d_n)}{d_n}$	*Optimal Return*	*Optimal Decision*
		$f_n^* (s_n)$	d_n^*

DYNAMIC PROGRAMMING UNDER CERTAINTY

Dynamic programming under certainty involves problems where the problem conditions at each stage, *i.e.,* state variables are known with certainty. In deterministic dynamic programming there is neither uncertainty nor probability distribution associated with the state in various stages of the decision process.

Model I. Shortest Route Problem

It is illustrated by some example

Model II. Single Additive Constraint, Multiplicative Separable Return

The general form of the recursive equation to solve such a problem by dynamic programming is illustrated by considering the following problem. Suppose we have separable return functions $f_j(d_j)$, $j = 1, 2, \ldots, n$ and desire to

$$\text{Maximize } Z = \{f_1(d_1).\ f_2(d_2), \ldots, f_n(d_n)\}$$

subject to the constraints

$$a_1d_1 + a_2d_2 + \ldots + a_nd_n = b$$

and $\quad d_j, a_j, b \geq 0$ for all $j = 1, 2, \ldots, n$

where $\quad j$ = jth number of stage $(j = 1, 2, \ldots, n)$

d_j = decision variable at jth stage; a_j = constant.

Let us now define state variables, $s_1, s_2, \ldots, s_n$ such that

$$s_n = a_1d_1 + a_2d_2 + \ldots + a_nd_n = b$$

$$s_{n-1} = a_1d_1 + a_2d_2 + \ldots + a_{n-1}d_{n-1} = s_n - a_nd_n$$

$$\vdots$$

$$s_{j-1} = s_j - a_jd_j$$

$$\vdots$$

$$s_1 = s_2 - a_2d_2$$

In general, it may be noted that the state transition function takes the form

$$s_{j-1} = t_j(s_j, d_j);\ j = 1, , \ldots, n,$$

i.e., a function of next state and decision variables.

At the nth stage, sn is expressed as the function of the decision variables. Thus the maximum value of Z denoted by $f_n^*(s_n)$ for any feasible value of sn is given by

$$f_n^*(s_n) = \underset{d_j > 0}{\text{Maximum}} \{f_1(d_1)\ ;\ f_2(d_2) \ldots f_n(d_n)$$

subject to the constraint $s_n = b$

Now for a moment holding a particular value of d_n fixed, the maximum value of Z will be given by

$$f_n(d_n) * \underset{d_j > 0}{\text{Max}} \{f_1(d_1) \,.\, f_2(d_2) \ldots f_{n-1}(d_{n-1})\};\ j = 1, 2, \ldots, n-1$$

$$= f_n(d_n) * f_{n-1}^*(s_{n-1})$$

The maximum value $f_{n-1}^*(d_{n-1})$ of Z due to decision variables d_j ($j = 1, 2, \ldots, n-1$) depends upon the state variable $s_{n-1} = t_n(s_n, d_n)$. The maximum of Z for any feasible value of all decision variables will be given by

$$f_j^*(s_j) = \underset{d_j > 0}{\text{Max}} \left[f_j(d_j) * f_{j-1}^*(s_j - 1)\right];\ j = n, n-1, \ldots, 2$$

$$f_1(s_1) = f_1(d_1)$$

where $s_{j-1} = t_j(s_j, d_j)$

The value of $f_j^*(s_j)$ represents the general recursive equation.

Model III. Single Additive Constraint, Additive Separable Return

Consider the problem

$$\text{Minimize } Z = [f_1(d_1) + f_2(d_2) + \ldots + f_n(d_n)]$$

subject to the constraint

$$a_1 d_1 + a_2 d_2 + \ldots + a_n d_n \geq b$$

and $a_j, d_j, b \geq 0$ for all j

Proceed in the same manner as in Model II. Defining state variables s_1, s_2, ... s_n such that

$$s_n = a_1 d_1 + a_2 d_2 + a_n d_n \geq b$$

$$s_{n-1} = s_n - a_n d_n$$

$$\vdots$$

$$s_{j-1} = s_j - a_j d_j;\ j = 1, 2, 3, \ldots, n$$

Let $$f_n^*(s_n) = \underset{d_j > 0}{\text{Min}} \sum_{n=1}^{n} f_j(d_j)$$

such that $s_n \geq b$

The general recursive equation for obtaining the minimum value of Z for all decision variables and for any feasible value of all decision variables is given by

$$f_j^*(s_j) = \underset{d_j > 0}{\text{Min}} \left[f_j(d_j) + f_{j-1}^*(s_{j-1})\right].\ j = 2, 3, \ldots, n$$

$$f_1^*(s_1) = f_1(d_1)$$

where $s_{j-1} = t_j(s_j, d_j)$

Mode IV. Single Multiplicative Constraint, Additively Separable Return

Consider the problem

$$\text{Minimize } Z = [f_1(d_1) + f_2(d_2) + \dots + f_n(d_n)]$$

subject to the constraint

$$d_1 \cdot d_2 \cdot d_3 \cdot \dots \cdot d_n \geq b$$

and $d_j, b \geq 0$ for all j

Proceed in the same manner as in Model II. Define state variables s_1, s_2, ... , s_n such that

$$s_n = d_n \cdot d_{n-1} \dots d_2 \cdot d_1 \geq b$$

$$s_{n-1} = d_{n-1} \cdot d_n - 2 \dots d_2 \cdot d_1 = s_n/d_n$$

$$\vdots$$

$$s_{j-1} = s_j/d_j;\ j = 2, 3, \dots, n$$

Let $f_n^*(s_n) = \underset{d_j > 0}{\text{Min}} \sum_{j=1}^{n} f_j(d_j)$ such that $s_n \geq b$

The general recursive equation for obtaining the minimum value of Z for all decision variables and for any feasible value of decision variable will be given by

$$f_j^*(s_j) = \underset{d_j > 0}{\text{Min}} [f_j(d_j) + f_{j-1}^*(s_{n-1})];\ j = 2, 3, \dots, n$$

$$f_j^*(s_1) = f_1(d_1)$$

and $s_{j-1} = t_j(s_j, d_j)$

DYNAMIC PROGRAMMING APPROACH FOR SOLVING LINEAR PROGRAMMING PROBLEM

A linear programming problem in n decision variables and m constraints can be converted into an n stage dynamic programming problem with m states. To illustrate, let us consider a general linear programming problem

$$\text{Maximize } Z = \sum_{j=1}^{n} c_j x_j$$

subject to the constraints

$$\sum_{j=1}^{n} a_{ij} x_j \leq b_i;\ i = 1, 2, \dots, m$$

and $x_j \geq 0;\ j = 1, 2, \ldots, n$

To solve an LP problem by using dynamic programming, the value of the decision variable x_j is determined at stage j (j = 1, 2, ... n). The value of xj at several stages can be obtained either by the forward or the backward induction method. The state variables at each stage are the amount of resources available for allocation to the current stage and succeeding stages.

Let $a_{1j}, a_{2j}, \ldots, a_{mj}$ be amount (in units) of resources i (i = 1, 2, ... ,m) respectively allocated to an activity c_j at jth stage, and fn $(a_{1j}, a_{2j}, \ldots, a_{mj})$ be the optimum value of the objective function of a general LP problem for stages j, j + 1, ... ,n, and for states $a_{1j}, a_{2j}, \ldots, a_{mj}$. Thus, the LP problem may be defined by a sequence of functions as:

$$f_n(a_{1j}, a_{2j}, \ldots, a_{mj}) = \text{Max} \sum_{j=1}^{n} c_j x_j$$

This maximization is taken over the decision variable x_j such that

$$\sum_{j=1}^{n} a_{ij} x_j \leq b_i;\ x_j \geq 0$$

The recursive relations for optimization are

$$f_n(a_{1n}, a_{2n}, \ldots, a_{mn}) \text{ or } f_n(b_1, b_2, \ldots, b_m)$$

$$= \underset{0 \leq a_{in} x_n \leq b_i}{\text{Max}} \{c_n x_n + f_{n-1}(b_1 - a_{1n} x_n, b_2 - a_{2n} x_n, \ldots, b_m - a_{mn} x_n)\}$$

The maximum value b that a variable x_n can assume is

$$b = \text{Min} \left\{ \frac{b_1}{a_{1n}}, \frac{b_2}{a_{2n}}, \ldots, \frac{b_m}{a_{mn}} \right\}$$

because the minimum value satisfies the set of constraints simultaneously.

SOLVED EXAMPLES

Example 1:

A salesman located in a city A decided to travel to city B. He knew the distances of alternative routes from city A to city B. He then drew a highway network map as shown in the Fig. 1.3. The city of origin, A, is city 1. The destination city B, is city 10. Other cities through which the salesman will have to pass through are numbered 2 to 9. The arrow representing routes between cities and distances in kilometers are indicated on each route. The salesman's problem is to find the shortest route that covers all the selected cities from A to B.

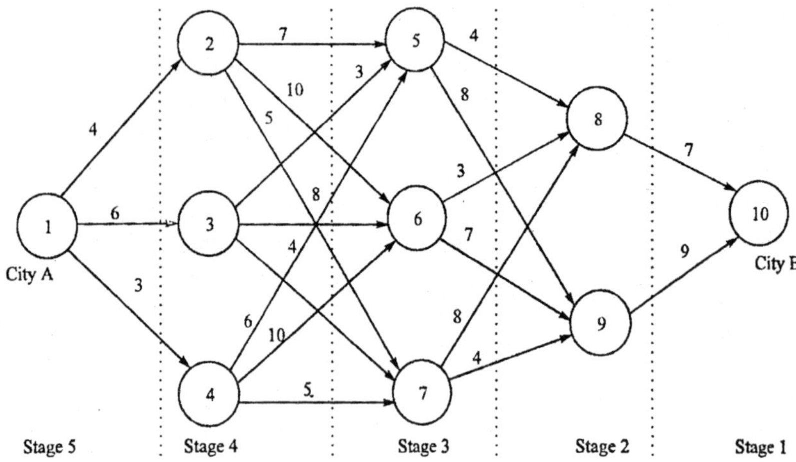

Fig. 1.3 : Network of Routes.

Solution:

To solve the problem, define problem stages, decision variables, state variables, return function and transition function. For this particular problem, the following definitions will be used to denote various state variables and decision variables.

d_n = decision variables that define the immediate destinations when there are n (n = 1, 2, 3, 4,) stages to go

s_n = state variables describe a specifies city at any stage

D_{sn}, d_n = distance associated with the state variable, sn, and the decision variables, dn for the current nth stage

$f_n (s_n, d_n)$ = minimum total distance for the last n stages, given that salesman is in state s_n and selects d_n as immediate destination

$f_n^* (s_n)$ =optimal path (minimum distance) when the salesman is in state s_n with n more stages to go for reaching the final stage (destination).

We start calculating distances between a pair of cities from destination city 10 (= x_1) and work backwards $x_5 \rightarrow x_4 \rightarrow x_3 \rightarrow x_2 \rightarrow x_1$ to find the optimal path. The recursion relationship for this problem can be stated as follows:

$$f_n^* (s_n) = \underset{d_n}{\text{Min}} \left\{ d_{s_n, d_n} + f_{n-1}^*(d_n) \right\}; \; n = 1, 2, 3, 4$$

where $f_{n-1}^* (d_n)$ is the optimal distance for the previous stages.

Working backward stages from city B to city A, we determine the shortest distance to city B (node 10) in stage 1, from state s_1 = 8 (node 8) and state s_1 = 9 (node 9) in stage 2. Since the distances associated with entering into stage 2 from state s_1 = 8 and s_1 = 9 are $_{D8,}$ 10 = 7 and $D_{9,10}$ = 9, respectively, the optimal value of $f_1^*(s_1)$ is the minimum value between $D_{8,10}$ and $D_{9,10}$. The results are shown in Table 1.2.

Table 1.2 : Stage 2

Decision, $d_1 \rightarrow$		$\frac{f_1(s_1,d_1) = D_{s_1,d_1}}{10}$ ***Distance*** $f_1^*(s_1)$	***Minimum Decision*** d_1	***Optimal***
States, s_1	8	7	7	10
9	9	9	10	

We more backward to stage 3. Suppose that the salesman is at state s_2=5 (node 5). Here he has to decide whether he should go to either d_2 = 8 (node 8) or d_2 = 9 (node 9). For this he must evaluate two sums

$$D_{5,8} + f_1^*(8) = 4 + 7 = 11 \text{ (to state } s_1 = 8)$$

$$D_{5,9} + f_1^*(9) = 8 + 9 = 17 \text{ (to state } s_1 = 9)$$

This distance function for travelling from state s_2 = 5 is the smallest of the set two sums:

$$f_2(s_2) = \underset{d_2 = 8,9}{\text{Min}} \{11, 17\} = 11 \text{ (to state } s_1 = 8)$$

Similarly, the calculation of distance function for travelling from state s_2 = 6 and s_2 = 7 can be completed as follows:

For state, s_2 = 6 $\quad f_2(6) = \underset{d_2 = 8,9}{\text{Min}} \begin{matrix} D_{6,8} + f_1^*(8) = 3 + 7 = 10 \\ D_{6,9} + f_1^*(9) = 7 + 9 = 16 \end{matrix}$

$$= 10 \text{ (to state } s_1 = 8)$$

For state, s_2 = 7 $\quad f_2(7) = \underset{d_2 = 8,9}{\text{Min}} \begin{matrix} D_{7,8} + f_1^*(8) = 8 + 7 = 15 \\ D_{7,9} + f_1^*(9) = 4 + 9 = 13 \end{matrix}$

$$= 13 \text{ (to state } s_1 = 9)$$

These results are entered into the two-stage table as shown in Table 1.3

Table 1.3 : Stage 3

Decision, d_2 →	$f_2(s_2, d_2) = D_{s2,d2} + f_1^*(d_2)$		Minimum Distance	Optimal Decision
	8	9	$f_2^*(s_2)$	d_2
States, s_2 5	11	17	11	8
6	10	16	10	8
7	15	13	13	9

Continuing the same process for stages 4 and 5, the results are shown in Tables 1.4 and 1.5.

Table 1.4 : Stage 4

Decision, d_3 →	$f_3(s_3, d_3) = D_{s3,d3} + f_2^*(d_3)$			Minimum Distance	Optimal Decision
	5	6	7	$f_3^*(s_3)$	d_3
States, s_3 2	18	20	18	18	5 or 7
3	14	18	17	14	5
4	17	20	18	17	5

Table 1.5 : Stage 5

Decision, d_4 →	$f_4(s_4, d_4) = D_{s4,d4} + f_3^*(d_4)$			Minimum Distance	Optimal Decision
	2	3	4	$f_4^*(s_4)$	d_4
States, s_4 1	2	20	20	20	3 or 4

The above optimal results at various stages can be summarized as given below:

Entering stages (nodes)

$$\text{Sequence} \begin{cases} 10 & 8 & 5 & 3 & 1 \\ 10 & 8 & 5 & 4 & 1 \end{cases}$$

$$\text{Distances} \begin{cases} 7 & 4 & 3 & 6 = 20 \\ 7 & 4 & 6 & 3 = 20 \end{cases}$$

From the above, it is clear that there are two alternative shortest routes for this problem, both having a minimum distance of 20 kilometers.

Example 2:

(Optimal sub-division problems) Divide quantity b into n parts so as to maximize their product. Let f_n (b) be the maximum value. Then show that

$$f_1(b) = b$$

$$f_n(b) = \underset{0 \le z \le b}{\text{Max}} \{n\, f_{n-1}(b - z)\}$$

Hence find f_n (b) and the division that maximize it.

Solution:

Let x_j be the jth part of the quantity b (j = 1, 2, ... ,n). Then the problem becomes Maximize $f_n(b) = x_1 . x_2 \ldots x_n$

subject to the constraints

$$x_1 + x_2 + \ldots + x_n = b$$

and $\quad x_j > 0;\ j = 1, 2, \ldots, n$

Here each part x_j (j = 1, 2, ... ,n) of b may be regarded as a stage. Since x_j may assume any positive value satisfying the given condition that $x_1 + x_2 + \ldots + x_n = b$, alternatives at each stage are infinite. Thus x_j's may be considered continuous variables. The recursive equation of the problem for all values of c can be obtained as follows:

For n = 1, the above result becomes f_1 (b) = x or b (initially true).

For n = 2 (*i.e.,* two stage problem), the quantity b is divided into two parts, say $x_1 = z$ and $x_2 = b - z$.

Then

$$f_2(b) = \text{Max}\,(x_1, x_2\} = \underset{0 < z \le b}{\text{Max}} \{z\,(b - z)\}$$

$$= \underset{0 < z \le b}{\text{Max}} \{z\, f_1(b - z)\}, \text{ since } f_1(b - z) = b - z$$

Similarly, for n = 3, the maximum product of b divided into three parts given the initial choice of z which leaves (b – z) to be further divided into two parts. Denoting the maximum possible product for (b – z) into two parts by f_2 (b – z). Thus, using the principle of optimally, we have

$$f_3(b) = \underset{0 < z \le b}{\text{Max}} \{z\, f_2(b - z)\}$$

Continuing in a similar manner, the recursive equation for general value of n is given by

$$f_n(b) = \underset{0 < z \le b}{\text{Max}} \{z\, f_{n-1}(b - z)\}$$

Solution to the recursive equation : The solution to recursive equation (1) to get the optimal policy can be obtained with the help of differential calculus.

For n = 2, the function z (b – z) attains its maximum value for z=b/2, satisfying the condition $0 < z \leq b$.

Hence, functional equation becomes

$$f_2(b) = \underset{0<z\leq b}{\text{Max}}\left\{\frac{b}{2}\cdot\left(b-\frac{b}{2}\right)\right\} = \left(\frac{b}{2}\right)^2$$

Then for n = 2, we have

$$\text{Optimal policy: } \left(\frac{b}{2},\frac{b}{2}\right) \text{ and } f_2(b) = \left(\frac{b}{2}\right)^2$$

For n = 3, the functional equations becomes

$$f_3(b) = \underset{0<z\leq b}{\text{Max}}\ \{z\ f_2(b-x)\} = \underset{0<z\leq b}{\text{Max}}\left\{z\left(\frac{v-z}{2}\right)^2\right\}$$

$$= \underset{0<z\leq b}{\text{Max}}\left\{z\left(\frac{b-z}{4}\right)^2\right\} = \left(\frac{b}{3}\right)^3$$

The maximum value of $z\left(\frac{b-z}{2}\right)^2$ is attained for $z = \frac{b}{3}$, satisfying the condition $0 < z \leq b$ because

$$f_2(b-z) = f_2\left(b-\frac{b}{3}\right) = f_2\left(\frac{2}{3}b\right) = \left\{\frac{1}{2}\left(\frac{2}{3}b\right)\right\}^2 = \left(\frac{1}{3}b\right)^2.$$

Thus for n = 3, we have

$$\text{Optimal policy: } \left(\frac{b}{3},\frac{b}{3},\frac{b}{3}\right)$$

and $$f_3(b) = \left(\frac{b}{3}\right)^2$$

Hence, in general, for an n-stage problem, we assume that

$$\text{Optimal policy: } \left(\frac{b}{n},\frac{b}{n},\ldots,\frac{b}{n}\right)$$

and $$f_n(b) = \left(\frac{b}{n}\right)^n \text{ form=1, 2, ... ,m.}$$

Now by induction it can also be shown that the result holds good for n = m = 1. The method is discussed below:

For n = m + 1, the functional equation becomes

$$f_{m+1}(b) = \underset{0 < z \le b}{\text{Max}} \{z\, f_m(b - z)\}$$

$$= \underset{0 < z \le b}{\text{Max}} \left\{ z\left(\frac{b-z}{m}\right)^m \right\}$$

$$= \underset{0 < z \le b}{\text{Max}} \left\{ z\frac{(b-z)^m}{(m)^m} \right\} = \left(\frac{b}{m+1}\right)^{m+1}$$

The maximum value of $z\left(\frac{b-z}{m}\right)^m$ is attained for $z = \frac{b}{m+1}$, that is, the result is true for n = m + 1 also.

Hence, the required optimal policy is

$$\left(\frac{b}{n}, \frac{b}{n}, \ldots, \frac{b}{n}\right) \text{ for } f_n(b) = \left(\frac{b}{n}\right)^n$$

Remark: The maximum value z $(b - z)^2$ was obtained by using the concept of maximum and minimum. For example, let

$$f(z) = \left\{\frac{b-z}{2}\right\}^2$$

Then $\quad \frac{d}{dz}\{f(z)\} = \frac{1}{4}\{2z(b-z)(-1) + (b-z)^2\}$

But for calculating maximum or minimum of f (z), we have two equate

$$\frac{d}{dz}\{f(z)\} = 0; \text{ i.e., } -2z(b - z) + (b - z)^2 = 0$$

This gives z = b/3. Further, the second derivative of f (z), *i.e.*, $\frac{d^2}{dz^2}\{f(z)\}$ is negative, at z = b/3. Hence the maximum value of f(z) is obtained at z = b/3.

Example 3:

Determine the value of u_1, u_2 and u_3 so as to

Maximize $Z = u_1 \cdot u_2 \cdot u_3$

subject to the constraint

$$u_1 + u_2 + u_3 = 10$$

and $\quad u_1, u_2, u_3 \ge 0.$

Solution:

Let us define state variable x_j (j = 1, 2, 3) such that

$x_3 = u_1 + u_2 + u_3 = 10$, at stage 3

$x_2 = x_3 - x_3 = u_1 + u_2$, at stage 2

$x_1 = x_2 - u_2 = u_1$, at stage 1

The maximum value of Z for any feasible value of state variable is given by

$$f_3(x_3) = \underset{u_3}{\text{Max}} \{u_3 \,.\, f_2(x_2)\}$$

$$f_2(x_2) = \underset{u_3}{\text{Max}} \{u_2 \,.\, f_1(x_1)\}$$

$$f_1(x_1) = u_1 = x_2 - u_2$$

Thus $f_2(x_2) = \underset{u_3}{\text{Max}} \{u_2 \,.\, (x_2 - u_2)\}$

$$= \underset{u_3}{\text{Max}} \{u_2 x_2 - u_2^2\}$$

Differentiating $f_2(x_2)$ with respect to u_2 and equating to zero (necessary condition for maximum or minimum value of a function), we have

$$x_2 - 2u_2 = 0 \text{ or } u_2 = x_2/2$$

Now using Bellman's principle of optimally, we get

$$f_2(x_2) = (x_2/_2).\, x_2 - (x_2/2)^2 = x_2^2/4$$

and $f_3(x_3) = \underset{u_3}{\text{Max}} \{u_3 \,.\, f_2(x_2)\} = \underset{u_3}{\text{Max}} \{u_3 \,.\, (x_2^2/4)\}$

$$= \text{Max} \left\{ u_3 . \frac{(x_3 - u_3)^2}{4} \right\}$$

Again differentiating $f_2(x_2)$ with respect to u_3 and equating to zero,

$$\frac{1}{4} \{u_3 \,.\, 2(x_3 - u_3)(-1) + (x_3 - u_3)^2\} = 0$$

$$(x_3 - u_3)(-2u_3 + x_3 - u_3) = 0$$

$$(x_3 - u_3)(x_3 - 3u_3) = 0$$

Now either $u_3 = x_3$ which is travails $u_1 + u_2 + u_3 = x_3$

or $x - 3u_3 = 0$ or $u_3 = x_3/3 = 10/3$. Therefore

$$u_2 = \frac{x_2}{2} = \frac{x_3 - u_3}{2} = \frac{1}{2}\left(10 \frac{10}{3}\right) = \frac{10}{3}$$

$$u_1 = x_2 - u_2 = \frac{20}{3} - \frac{10}{3} = \frac{10}{3}$$

Thus $u_1 = u_2 = u_3 = 10/3$ and hence Max $\{u_1 . u_2 . u_3\}$

$= (10/3)^3 = 1000/27.$

Example 4:

A company has decided to introduce a product in three phases. Phase 1 will feature making a special offer at a greatly reduce d rate to attract the first-time buyers. Phase 2 will involve intensive advertising to persuade the buyers to continue purchasing at a regular price. Phase 3 will involve a follow up advertising and promotional campaign.

A total of Rs. 5 million has been budgeted for this marketing campaign. If miss the market share captured in Phase 1, fraction f_2 of m is retained in Phase 2, and fraction f_3 of market share in Phase 2 is retained in Phase 3. The expected values of m, f_2 and f_3 at different levels of money expended are given below. How should the money be allocated to the three phases to maximize the final share?

Money	***Effect on Market Share***		
(Rs millions)	***m per cent***	***f2***	***f3***
0	*0*	*0.30*	*0.50*
1	*10*	*0.50*	*0.70*
2	*15*	*0.70*	*0.85*
3	*22*	*0.80*	*0.90*
4	*27*	*0.85*	*0.93*
5	*30*	*0.90*	*0.95*

Solution:

This problem can be treated as a three-stage problem taking each phase as a stage and amount of money spent as the state of the system. Let us adopt the following notations:

s = millions of rupees spent on marketing campaign at a particular stag

x_j = amount of money allocated to phase (stage) j; (j = 1, 2, 3)

$P_j(x_j)$ = market share captured in phase j; (j = 1, 2, 3)

The problem can now be expressed mathematically as:

Maximize $Z = p_1 (x_1) \cdot p_2 (x_2) \cdot p_3 (x_3)$

subject to the constraint

$$x_1 + x_2 + x_3 = 5$$

and $x_1, x_2, x_3 \geq 0$

Using the forward induction approach, the recursive equation calculations are as follows

$$f_j^* (s) = \max_{0 \leq x_j \leq s} \{p_j (s_n) \times f_{j-1}^* (s - x_j)\}$$

Phase 1 (j = 1). If the whole amount is spent in Phase 1, that is $x_1 \leq 5$, the optimal return will be

x_1^*	:	0	1	2	3	4	5
$m = f_1^*(x_1)$	:	0	10	15	22	27	30

Phase 2 (j = 2): If the whole amount is spent in Phase 2, then

$$f_2^* (s) = \max_{0 \leq x_2 \leq s} \{f_2 (x_2) \times f_1^* (s - x_2)\}$$

$$= \max_{0 \leq x_2 \leq 5} \{f_2 (x_2) \times f_1^* (5 - x_2)\}$$

The computations for stages 2 and 3 are shown in Table 1.6 and 1.7, respectively.

Table 1.6 : Phase 2 (j = 2)

States s	*Decision* x_2 0	$f_2(s, x_2) = f_2(x_2) \times f_1^*(s - x_2)$ 1	2	3	4	5	*Optimal Return* f_2^* *(s)*	*Optimal Decision* x_2^*
0	0	–	–	–	–	–	0	0
1	3.0	–	–	–	–	–	3.0	0
2	4.5	5.0	–	–	–	–	5.0	1
3	6.6	7.5	7.0	–	–	–	7.5	1
4	8.1	11.0	10.5	8.0	–	–	11.0	1
5	9.0	13.5	15.4	12.0	8.5	–	15.4	2

Table 1.7 : Phase 3 (j = 3)

	Decision	$f_3(s, x_3) = f_3(x_3) \times f_2(s - x_3)$					*Optimal*	*Optimal*
	x_3						*Return*	*Decision*
States s	*0*	*1*	*2*	*3*	*4*	*5*	f_3^* (s)	x_3^*
5	7.7	7.7	6.37	4.5	2.79	–	7.7	0, 1

In Table 1.7, when $x_3 = 0$,

$s = x_1 + x_2 = 5$,

and Max f_2^* (s) = 15.4 for which $x_2 = 2$ and,

therefore, $x_1 = 3$.

But for $x_3 = 1$, $s = x_1 + x_2 = 4$,

and Max f_2^* (s) = 11.0 for which $x_2 = 1$ and,

therefore, $x_1 = 3$.

Hence, the optimal policy to obtain a maximum market share of 7.7 per cent can be any one of the following:

(i) spend 3, 2, 0 million rupees in first, second and third phases (stage), respectively, or

(ii) spend 3, 1, 1 million rupees in first, second and third phases (stage), respectively

Example 5:

An electronic device consists of four components each of which must function for the system to function. The system reliability can be improved by installing parallel units in one or more components. The reliability of components, R with one, two, or three parallel units, and the corresponding cost, C are given below. The maximum amount available for this device is 100. The problem is to determine the number of parallel units in each component.

Number of	Components							
Parallel Units	*1*		*2*		*3*		*4*	
	R	C	R	C	R	C	R	C
1	*0.70*	*10*	*0.50*	*20*	*0.70*	*10*	*0.60*	*20*
2	*0.80*	*20*	*0.70*	*40*	*0.90*	*30*	*0.70*	*30*
3	*0.90*	*30*	*0.80*	*50*	*0.95*	*40*	*0.90*	*40*

Solution:

The reliability of the given electronic device is the product of the reliabilities of each of its four components. If R_j and u_j represents the reliability of the component j and units in parallel in component i, then the reliability of whole system consists of n components in series will be:

$$R_1u_1 \times R_2u_2 \times ... \times R_nu_n.$$

Since the objective is to maximize the reliability of the system, the problem can be staged as:

$$\text{Maximize } Z = R_1u_1 \times R_2u_2 \times ... \times R_nu_n$$

subject to the constraint

$$c_1u_1 + c_2u_2 + ... + c_nu_n \leq C$$

where $c_j\ u_j$ = cost of component j

C = total capital available

Since the electronic device consists of n = 4 components, for solving this problem consider each component as a stage. The state at any stage will be the capital to be allocated. Let us adopt the following notations.

x_j = capital allocated to stage j, through first stage inclusive.

$f_j\ (x_j)$ = return when available capital cj is allocated optimally over n stages (components)

R_ju_j = reliability of component j (j = 1, 2, 3, 4)

Now the recursive equation can be expressed as:

$$f_j\ (x_j) = \underset{0 < c_ju_j < x_j}{\text{Max}} [\{R_1u_1 \times R_2u_2 \times ... \times R_nu_n\}.\ f_{j-1}\ (x_j - c_j\ u_j];$$

$$j = 1, 2, 3, 4.$$

The electronic device in this problem will consist of at least one unit in each component. Thus, the range of investment, xj (j = 1, 2, 3, 4) in each case will be as follow:

$$c_{11} \leq x_1 \leq C - c_{21} - c_{31} - c_{41} \text{ or } 10 \leq x_1 \leq 50$$

$$c_{11} + c_{12} \leq x_2 \leq C - c_{31} - c_{41} \text{ or } 30 \leq x_2 \leq 70$$

$$c_{11} + c_{12} + c_{13} \leq x_3 \leq C - c_{41} \text{ or } 40 \leq x_3 \leq 80$$

$$c_{11} + c_{12} + c_{12} + c_{14} \leq x_4 \leq C \text{ or } 60 \leq x_4 \leq 100$$

The computation of return at each stage using the forward induction approach, is shown in the following tables.

Table 1.8 : Stage 1 (j = 1)

States	Decision u_1 $f_1(x_1) = R_1u_1$ $u_1 = 1$ / $R = 0.70; C = 10$	$u_1 = 2$ / $R = 0.80; C = 20$	$u_1 = 3$ / $R = 0.90; C = 30$	Optimal Return $f_1^*(x_1)$	Optimal Decision u_1^*
10	0.70	–	–	0.70	1
20	0.70	0.80	–	0.80	2
30	0.70	0.80	0.90	0.90	3
40	0.70	0.80	0.90	0.90	3
50	0.70	0.80	0.90	0.90	3

Table 1.9 : Stage 2 (j = 2)

States	Decision u_2 $f_2(x_2) = R_2u_2 \times f_1^*(x_2 - c_2x_2)$ $u_2 = 1$ / $R = 0.50; C = 20$	$u_2 = 2$ / $R = 0.70; C = 40$	$u_2 = 3$ / $R = 0.80; C = 50$	Optimal Return $f_2^*(x_2)$	Optimal Decision u_2^*
30	0.5×0.7 = 0.35	–	–	0.35	1
40	0.5×0.8 = 0.40	–	–	0.40	1
50	0.5×0.9 = 0.45	0.7×0.7 = 0.49	–	0.49	2
60	0.5×0.9 = 0.45	0.7×0.8 = 0.56	0.8×0.7. = 0.56	0.56	2, 3
70	0.5×0.9 = 0.45	0.7×0.9 = 0.63	0.8×0.8 = 0.64	0.64	3

Table 1.10 : Stage 3 (j = 3)

States x_3	Decision u_3 $f_3(x_3) = R_3u_3 \times f_2^*(x_3 - c_3x_3)$ $u_3 = 1$ / $R = 0.70; C = 10$	$u_3 = 3$ / $R = 0.90; C = 40$	$u_3 = 3$ / $R = 0.90; C = 40$	Optimal Return $f_3^*(x_3)$	Optimal Decision u_3^*
40	0.7×0.35=0.245	–	–	0.245	1
50	0.7×0.40=0.280	–	–	0.280	1
60	0.7×0.49=0.343	0.9×0.35=0.315	–	0.343	1
70	0.7×0.56=0.392	0.9×0.40=0.360	0.95×0.35=0.3325	0.392	1
80	0.7×0.64=0.448	0.9×0.49=0.441	0.95×0.40=0.3800	0.448	1

Table 1.11 : Stage 4 (j = 4)

States x_4	Decision u_4 / $f_4(x_4) = R_4u_4 \times f_3^*(x_4 - c_4u_4)$ $u_4 = 1$, R = 0.60; C = 20	$u_4 = 2$, R = 0.70; C = 30	$u_4 = 3$, R = 0.90; C = 40	Optimal Return $f4^*(x4)$	Optimal Decision u_4^*
60	0.6×0.245=0.147	–	–	0.147	1
70	0.6×0.280=0.168	0.7×0.245=0.171	–	0.171	2
80	0.6×0.343=0.205	0.7×0.280=0.196	0.9×0.245=0.220	0.220	3
90	0.6×0.392=0.235	0.7×0.343=0.240	0.9×0.280=0.252	0.252	3
100	0.6×0.448=0.268	0.7×0.392=0.274	0.9×0.343=0.308	0.308	3

In Table 1.11, at stage 4 the value of return function $f_4(x_4)$ is maximum, *i.e.*, 0.308 at $x_4 = 100$, and $u_4 = 3$. This makes $x_3 = 100 - 40 = 60$. In Table 1.10, $x_3 = 60$ corresponds to $u_3 = 1$, and we are left with $x_2 = 60 - 10 = 50$. In Table 1.9, $x_2 = 50$ corresponds to $u_2 = 2$, and we are left with $x_1 = 50 - 40 = 10$. In Table 1.8, $x_1 = 10$ corresponds to $u_1 = 1$.

Hence for maximum reliability, the device must have 1, 2, 1 and 3 units in components 1, 2, 3 and 4, respectively to attain a maximum reliability of 0.308 or 30.8 per cent.

Example 6:

Consider the problem of designing electronic devices to carry five power cells, each of which must be located within three electronic systems. If one system's power fails, then it will be powered on an auxiliary basis by the cells of the remaining systems. The probability that any particular system will experience a power failure depends on the number of cells originally assigned to it. Estimated power failure probabilities for a particular system are given below:

Power Cells	*Probability of System Power Failure* System 1	*System 2*	*System 3*
1	*0.50*	*0.60*	*0.40*
2	*0.15*	*0.20*	*0.25*
3	*0.04*	*0.10*	*0.10*
4	*0.02*	*0.05*	*0.05*
5	*0.01*	*0.02*	*0.01*

Determine how many power cells should be assigned to each system to maximize the overall system reliability.

Solution:

Let us adopt the following notations:

x_n = number of power cells assigned to stage

$p_n(x_n)$ = probability of power failure for the system n, when it is assigned x_n power cells

$f_n(s)$ = probability that nth and all higher systems will fail, while entering with state s

Here stages correspond to systems and state s is the number of power cells available for allocation at different stages. We shall start from state (power cell) 1. The recursive equations for this problem may be given by

$$f_n(s) = \min_{x_n \leq 5} \{p_n(x_n) \times f_{n+1}(s - x_n)\},\ n = 1, 2$$

subject to the constraint

$$x_1 + x_2 + \dots + x_n = 5$$

The dynamic programming calculations are as follows:

Table 1.12 : Stage 3 (n = 3); $f_3(s) = \min_{x_3 \leq 5} \{p_3(x_3)\}$

Decision x_3 / *States, s*	*$p_3(x_3)$* 1	2	3	*Minimum Value* $f_3^*(s)$	*Optimal Decision* x_3^*
1	0.40	–	–	0.40	1
2	0.40	0.25	–	0.25	2
3	0.40	0.25	0.10	0.10	3

Table 1.13 : Stage 2 (n = 2)

Decision x_2 / *States, s*	$f_2(s) = p_2(x_2) \times f_s^*(s - x_2)$ 1	2	3	*Minimum Value* $f_2^*(s)$	*Optimal Decision* x_2^*
2	0.24	–	–	0.24	1
3	0.15	0.08	–	0.08	2
4	0.06	0.05	0.04	0.04	3

Table 1.14 Stage 1 (n = 1)

Section x_1	$f_1(s) = p_1(x_1) \times f_2^*(s - x_1)$			Minimum Value	Optimal Decision
States, s	1	2	3	$f_1^*(s)$	x_1^*
5	0.2	0.012	0.0096	0.0096	3

At stage 1, value of $f_1(s) = 0.0096$ is minimum when $x_1 = 3$ corresponds to system 1. Then for $x_1 = 3$, $x_2 + x_3 = 2$. But at stages 2 and 1, optimal values of x_2 and x_3 are 1 and 1, respectively. Thus optimal solution is: $x_1 = 3$, $x_2 = 1$, $x_3 = 1$ with the smallest probability of total power failure, $f_1(5) = 0.0096$.

Example 7:

Use dynamic programming to show that

$$\underset{d_j > 0}{\text{Min}} = 1$$

and $p_i \geq 0$, *for all i*

is minimum when $p_1 = p_2 = \ldots = p_n = 1/n$.

Solution:

The problem is to divide unity into n parts, $p_1, p_2, \ldots, p_n$ such that the quantity

$$f_n(1) = \sum_{i=1}^{n} p_i \log p_i$$

is minimum, where $f_n(1)$ denotes the minimum attainable sum of $p_i \log p_i$ $(i = 1, 2, \ldots, n)$, when 1 is divided into n parts.

For n = 1 (Stage 1) we have,

$f_1(1) = p_1 \log p_1 = 1 \log 1$, as unity is divided only into $p_1 = 1$ part.

For n = 2 (Stage 2) the unity is to be divided into two parts p_1 and p_2 such that $p_1 + p_2 = 1$. If $p_1 = z$ and $p_2 = 1 - z$, then

$$f_2(1) = \underset{0 < z \leq 1}{\text{Min}} \{p_1 \log p_1 + p_2 \log p_2\}$$

$$= \underset{0 < z \leq 1}{\text{Min}} \{z \log z + (1 - z) \log (1 - z)\}$$

$$= \underset{0 < z \leq 1}{\text{Min}} \{z \log z + f_1(1 - z)\}$$

Similarly, in general, for an n-stage problem, where unity is to be divided into on parts, the recursive equations is

$$f_n(1) = \underset{0 < z \leq 1}{\text{Min}} \{p_1 \log p_1 + p_2 \log p_2 + \ldots + p_n \log p_n\}$$

$$= \underset{0 < z \leq 1}{\text{Min}} \{z \log z + f_{n-1}(1 - z)\}$$

Solution of recursive equation : The solution to the above recursive equation to get the optimal policy can be obtained with the help of differential calculus.

The minimum value of recursive equation

$$f(z) = z \log(z) + (1 - z) \log(1 - z)$$

is attained at $z = 1/2$, satisfying the condition $0 < z \leq 1$. Hence, for stage 2, the optimal policy is $p_1 = p_2 = 1/2$ and

$$f_2(1) = \frac{1}{2}\log\frac{1}{2} + \left(1 - \frac{1}{2}\right)\log\left(1 - \frac{1}{2}\right) = 2\left(\frac{1}{2}\log\frac{1}{2}\right)$$

Similarly, for stage 3, the minimum value of the recursive equation

$$f_3(1) = \underset{0 < z \leq 1}{\text{Min}} \{z \log z + f_2(1 - z)$$

$$= \underset{0 < z \leq 1}{\text{Min}} \left\{z \log z + 2\left(\frac{1 - z}{2}\right)\log\left(\frac{1 - z}{2}\right)\right\}$$

is attained at $z = 1/3$ satisfying the condition $0 < z \leq 1$. Hence, for stage 3, the optimal policy is, $p_1 = p_2 = 1/3$ and

$$f_3(1) = \frac{1}{3}\log\left(\frac{1}{3}\right) + 2\left(\frac{1 - \frac{1}{3}}{2}\right)\log\left(\frac{1 - \frac{1}{3}}{2}\right)$$

$$= \frac{1}{3}\log\left(\frac{1}{3}\right) + 2\left\{\frac{1}{3}\log\left(\frac{1}{3}\right)\right\} = 3.\left\{\frac{1}{3}\log\left(\frac{1}{3}\right)\right\}$$

Thus, in general, the optimal policy is

$$p_1 = p_2 = \ldots = p_n = 1/n$$

and $$f_n(1) = n\left\{\frac{1}{n}\log\left(\frac{1}{n}\right)\right\}$$

The above result can be proved valid for any value of n by using mathematical induction. For $n = m + 1$, the recursive equation is

$$f_{m+1}(1) = \underset{0 < z \leq 1}{\text{Min}} \{z \log z + f_m (1 - z)\}$$

Since the minimum of the function

$$F(z) = z \log z + \frac{1-z}{m} \log\left(\frac{1-z}{m}\right)$$

is attained for $z = 1/(x + 1)$, therefore using the concept of maximum and minimum, we have

$$f_{m+1}(1) = \frac{1}{m+1} \log\left(\frac{1}{m+1}\right) + m\left\{\frac{1}{m+1} \log\left(\frac{1}{m+1}\right)\right\}$$

$$= (m + 1) \left\{\frac{1}{m+1} \log\left(\frac{1}{m+1}\right)\right\}$$

Hence, the result is true for $n = m + 1$ also.

Example 8:

Use dynamic programming to solve the following problem:

Minimize $Z = y_1^2 + y_2^2 + y_3^2$

subject to the constraint

$$y_1 + y_2 + y_3 = 10$$

and $y_1, y_2, y_3 \geq 0$

Solution:

Let the state variables be s_1, s_2 and s_3 such that

$$s_3 = y_1 + y_2 + y_3 = 15$$

$$s_2 = y_1 + y_2 = s_3 - y_3$$

$$s_1 = y_1 = s_2 - y_2$$

The recursive equations can now be written as:

$$f_3(s_3) = \underset{y_3}{\text{Min}} \{y_3^2 + f_2(s_2)\}$$

$$f_2(s_2) = \underset{y_2}{\text{Min}} \{y_2^2 + f_1(s_1)\}$$

$$f_1(s_1) = \underset{y_1}{\text{Min}} \{y_1^2\} = y_1^2 = (s_2 - y_2)^2$$

The recursive equation $f_2(s_2)$ can also be expressed as:

$$f_2(s_2) = \underset{y_2}{\text{Min}} \{y_2^2 (s_2 - y_2)^2\};\ f_1(s_1) = (s_2 - y_2)^2$$

Now by using the concept of maxima and minima in differential calculus, the minimum value of $f_2(s_2)$ can be obtained as explained below:

Differentiating $f_2(s_2)$ with respect to y_2 and equating to zero, we get

$$2y_2 - 2(s_2 - y_2) = \text{or } y_2 = s_2/2$$

Thus $$f_1(s_2) = \left(\frac{s_2}{2}\right)^2 + \left\{s_2 - \frac{s_2}{2}\right\}^2 = \frac{s_2^2}{2}$$

$$f_3(s_3) = \underset{y_3}{\text{Min}} [y_3^2 + f_2(s_2)\} = \underset{y_3}{\text{Min}} \left\{y_3^2 + \frac{s_2^2}{2}\right\}$$

$$= \underset{y_3}{\text{Min}} \left\{y_3^2 + \frac{(s_2 - y_3)^2}{2}\right\}$$

Differentiating $f_3(s_3)$ with respect to y_3 and equating to zero to get its minimum value,

$$2y_3 - (s_3 - y_3) = 0 \text{ or } y_3 - s_3/3$$

Thus $$f_3(s_3) = \left(\frac{s_3}{3}\right)^2 + \frac{1}{2}\left(s_3 - \frac{s_3}{3}\right)^3 = \frac{s_3^2}{3}.$$

But for minimum value of $f_3(s_3)$ we must have $y_1 + y_2 + y_3 = 10$. Therefore

$f_3(s_3) = (10)2/3 = 100/3$, and

$y_3 = s_3/3 = 10/3$

$y_2 = s_2/2 = (s_3 - y_3)/2 = \{10 - (10/3)\}/2 = 10/3$

$y_1 = s_1/2 = (s_3 - y_3) - y_2 = \{10 - (10/3)\{ - 10/3 = 10/3$

Hence, the minimum value of $y_1^2 + y_2^2 + y_3^2$ is 100/3

for $y_1 = y_2 = y_3 = 10/3$.

Example 9:

Suppose there are n machines which can perform two jobs. If x of them do the first job, then they produce goods worth g(x) = 3x, and if y of them perform the second job, then they produce goods worth h(y) = 2.5y. Machines are subject to depreciation so that after performing the first job only a(x) = x/3 machines remain available and after perforing the second job b(y) = 2/3 machines remain available in the beginning of the second year. The process is repeated with remaining machines. Obtain the maximum total return after three years and also find the optimal policy in each year.

Solution:

Let us define the following notations

n = stages, each year being viewed as a stage; $n = 1, 2, 3$

x_n = number of machines devoted to job 1 in the year n

y_n = number of machines devoted to job 2 in the year n

s = state variable, the number of machines at any stage (year)

$f_j(s)$ = maximum return function when initial available machines are s with n more stages (years) to go

Stage 1 (n = 1) In this problem, there are three stages (years). Using the backward induction approach, *i.e.*, $n = 1$ (third year), we have $s = s_3$ number of machines at the beginning of the year. Then the return function can be expressed as:

$$f_1(s_3) = \underset{x_3, y_3}{\text{Max}} \{3x_3 + 2.5y_3\}$$

subject to the constraint

$$x_3 + y_3 \le s_3 \quad \text{and} \quad x_3, y_3 \ge 0$$

where x_3, y_3 = number of machines devoted to jobs 1 and 2, respectively in the third year.

Since function $f_1(s_3)$ is a linear function in x_3 and y_3, its maximum can be obtained by using the concept of extreme point solutions of LP problem as shown in Fig. 1.4.

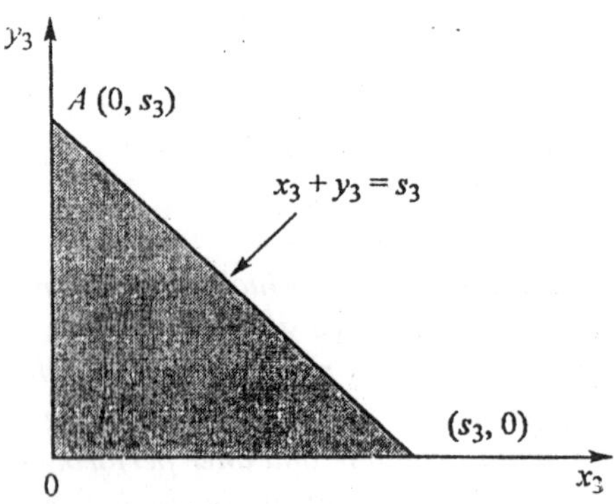

Fig. 1.4 : Extreme Point Solution (Stage 1).

As shown in Fig. 1.4, the maximum value of return function occurs at B $(s_3, 0)$, where

$$f_1(s_3) = 3x_3 + 2.5y_3$$
$$= 3s_3 + 2.5 \times 0 = 3s_3.$$

Hence optimal decision at this stage is: $x_3^* = s_3$, $y_3^* = 0$

and $f_1^*(s_3) = 3s_3$.

Stage 2 (n = 2) : Consider now the second year as the second stage. The return function at this stage is

$$f_2(s_2) = \max\begin{Bmatrix}\text{Immediate return} & \text{Maximum known return} \\ \text{from stage +} & \text{from stage 1}\end{Bmatrix}$$

$$= \underset{x_2, y_2}{\text{Max}} \left\{(3x_2 + 2.5y_2) + f_1^*\left(\frac{x_2}{3} + \frac{2y_2}{3}\right)\right\} y_2$$

subject to the constraint

$x_2 + y_2 \le s_2$ and $x_2, y_2 \ge 0$

where x_2 and y_2 = number of machines devoted to jobs 1 and 2, respectively in the second year.

$x_2/3$ and $2y_2/3$ = machines which will be available at the beginning of the next year.

By the definition of $f_1^*(s_3) = 3s_3$ (Stage 1) the return function of this stage becomes

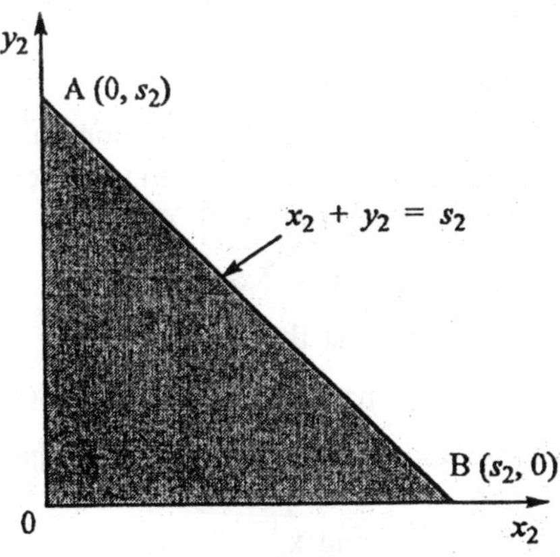

Fig. 1.5 : Extreme Point Solution (Stage 2).

$$f_2(s_2) = \underset{x_2, y_2}{\text{Max}} \left\{3x_2 + 2.5.y_2 + 3\left(\frac{x_2}{3} + \frac{2y_2}{3}\right)\right\}$$

$$= \underset{x_2, y_2}{\text{Max}} \{4x_2 + 4.5y_2\}$$

This is again a linear function in x_2 and y_2, and its maximum value occurs at corner point A $(0, s_2)$ as shown in Fig. 1.5.

The maximum value of the return function at A $(0, s_2)$ is: $f_2(s_2) = 4 \times 0 + 4.5 \times s_2 = 4.5s_2$. Hence, the optimal decision at this stage is: $x_2^* = 0$, $y_2^* = s_2$ and $f_2^*(s_2) = 4.5s_2$

Stage 3 (n = 3) Consider now the first year as the third stage. The return function at this stage is

$$f_3(s_1) = \left\{\begin{matrix} \text{Immediate return} & \text{Maximum known return} \\ \text{from stage } 3 + & \text{from stage } 2 \end{matrix}\right\}$$

$$= \underset{x_1, y_1}{\text{Max}} \left\{(3x_1 + 2.5y_1) + f_2^*\left(\frac{x_1}{3} + \frac{2y_1}{3}\right)\right\}$$

$$= \underset{x_1, y_1}{\text{Max}} \left\{3x_1 + 2.5y_1 + 4.5\left(\frac{x_1}{3} + \frac{2y_1}{3}\right)\right\} \quad f_2(s_2) = 4.5s$$

$$= \underset{x_1, y_1}{\text{Max}} \{4.5x_1 + 5.5y_1\}$$

subject to the constraint

$$x_1 + y_1 \leq s_1$$

and $\quad x_1, y_1 \geq 0$

This return function is also a linear function in x_1 and y_1, and its maximum value occurs at corner point A (0, s1) as shown in Fig. 1.6. The maximum value of return function at A (0, s1) is

$$f_3(s_1) = 4.5 \times 0 + 5.5 \times s_1 = 5.5s_1$$

Hence the optimal decision at this stage is

$$x_1^* = 0, y_1^* = s_1 \text{ (= m machines, say) and } f_3^*(s_1) = 5.5m$$

The values of x_2, x_3 and y_2, y_2 can also be obtained in terms of m as follows:

$$y_2 = s_2 = 2y_1/3 = 2m/3 \text{ and } x_2 = 0$$

$$x_3 = s_3 = 2y_2/3 = 2/3\ (2m/3) = 4m/9, \text{ and } y_3 = 0$$

Table 1.15 summarizes the optimal decision at each stage (year).

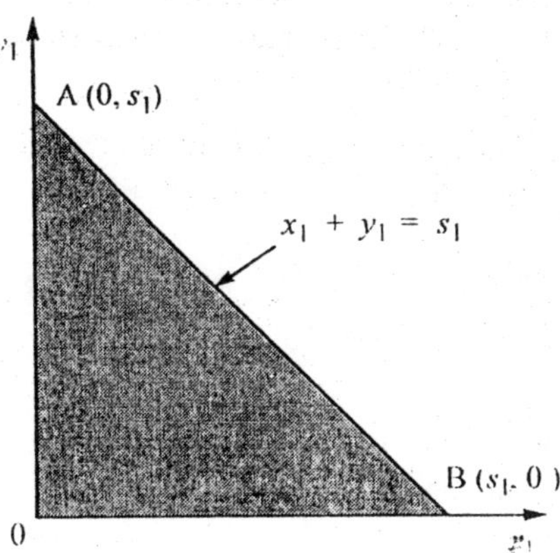

Fig 1.6 : Extreme Point Solution (Stage 3).

Table 1.5 : Summary of the Optimal Decisions

Stage	*Values of Decision variable*
3 (First year)	$x_1 = 0,\ y_1 = m$
2 (Second year)	$x_2 = 0,\ y_2 = 2m/3$
1 (Third year)	$x_3 = 4x/9,\ y_3 = 0$

The maximum possible return is 5.5m.

Example 10:

A man is engaged in buying and selling identical items, he operates from a warehouse having a capacity of 500 items. Each month he can sell any quantity that he chooses up to the stock at the beginning of the month. Each month, he can buy as much as he wishes for delivery at the end of the month so long as his stock does not exceed 500 items. For the next four months he has the following error free forecasts of cost and sales prices:

Month n	:	*1*	*2*	*3*	*4*
Cost, c_n	:	*27*	*24*	*26*	*28*
Sales price, p_n	:	*28*	*25*	*25*	*27*

If he currently has a stock of 200 units, what quantities should he and buy in the next four months? Find the solution using dynamic programming.

Solution:

Let us adopt the following notations:

x_n = amount to be purchased during month, n

y_n = amount to be sold during month, n

s_n = starting inventory for each month, n (state variable)

p_n = sales price in month, n

c_n = purchase price in month, n

$f_n(s_n)$ = optimal total return from the next n stages if stage n begins with inventory level sn.

To solve this problem we will use backward induction approach, *i.e.,* month 4 be stage 1 and month 1 be stage 4 so that the stage number measures the future time remaining.

The state of the warehouse is the starting inventory s_n for each month. The state transformation therefore, given by

$$s_{n-1} = \text{Inventory at start of month } n - 1$$

$$= s_n + x_n - y_n$$

Also, since selling precedes buying, the man cannot sell more than he owns at the start of the month: that is, $y_n \leq s_n$

The ending inventory for the month cannot exceed the warehouse capacity. Therefore

$$x_n + x_n - y_n \leq 500$$

or Purchase order, $x_n \leq 500 - s_n + y_n$

The return from each month's buying minus selling is the net revenue, *i.e.,*

Revenue, $r_n = y_n P_n - x_n c_n$

Let $f_n(s_n)$ represent the maximum return from the stage n, if it begins with inventory level s_n. Then recursive equation becomes

$$f_n^*(s_n) = \text{Max} \{y_n p_n - x_n c_n + f_{n-1}^*(s_n + x_n - y_n)\}$$

where $0 \leq y_n \leq s_n$

$0 \leq x_n \leq 500 - s_n + y_n$

Stage 1 (n = 1) : Since this sub-problem has no future to follow it, the return function is given by

$$f_1^*(s_1) = \text{Max}(27y_1 - 28x_1)$$

where $0 \leq y_1 \leq s_1$

$0 \leq x_1 \leq 500 - s_1 + y_1$

This represent a linear programming problem with two variables. The solution of it can be obtained graphically. The value of the objective function increases with the value of y_1 and decreases with value of x1. Therefore, the optimal solution is: $x_1 = 0$, $y_1 = s_1$ with the maximum return

$$f_1^* (s_1) = p_1y_1 = 27s_1$$

and $s_1 = s_2 + x_2 - y_2$

Stage 2 (n = 2) : $f_2^* (s_2) = \text{Max } \{25y_2 - 26x_2 + f_1^* (s_2 + x_2 - y_2)\}$

where $0 \leq y_2 \leq s_2$

$0 \leq x_2 \leq 500 = s_2 + y_2$

Since $f_1^* (s_1) = 27s_1$, substituting this value in the expression for $f_2 (s_2)$, we have

$$f_2^* (s_2) = \text{Max } \{25y_2 - 26x_2 + 27 (s_2 + x_2 - y_2)\}$$

$$= \text{Max } \{27s_2 + x_2 - 2y_2\}$$

The optimal solution at this stage can be obtained by comparing the following alternative solutions (corners of the feasible solution space).

(i) $x_2 = 0$, $y_2 = 0$, $f_2 (s2) = 27s_2$

(ii) $x_2 = 0$, $y_2 = s_2$, $f_2 (s2) = 25s_2$

(iii) $x_2 = 500$, $y_2 = s_2$, $f_2 (s_2) = 25s_2 + 500$

(iv) $x_2 = 500 - s_2$, $y_2 = 0$, $f_2 (s_2) = 26s_2 + 500$

Hence for any value of $s_2 \leq 500$, the optimal return is

$$f_2^* (s_2) = 26s_2 + 500; \; s_2 = s_3 + x_3 - y_3$$

for $x_2 = 500 - s_2$ and $y_2 = 0$

Stage 3 (n = 3) : $f_3^* (s_3) = \text{Max } \{25y_3 - 24x_3 + f_3^* (s_3 + x_3 - y_3)\}$

$= \text{Max } \{25y_3 - 24x_3 + 26 (s_3 + x_3 - y_3) + 500\}$

$= \text{Max } \{26s_3 + 2x_3 - y_3 + 500\}$ the optimal value of $f_3 (s_3)$ can be obtained by comparing the following alternative solution:

(i) $x_3 = 0$, $y_3 = 0$, $f_3 (s_3 = 26s_3 + 500$

(ii) $x_3 = 0$, $y_3 = s_2$, $f_3 (s_3 = 25s_3 + 500$

(iii) $x_3 = 500$, $y_3 = s_2$, $f_3 (s_3) = 25s_3 + 1{,}500$

(iv) $x_3 = 500 - s_3$, $y_3 = 0$, $f_2(s_2) = 24s_3 + 1{,}500$

Hence for any value of $s_3 \leq 500$, the optimal return is

$$f_3^*(s_3) = 25s_3 + 1{,}500;\; s_3 = s_4 + x_4 - y_4$$

for $x_3 = 500$ and $y_2 = s_3$

Stage 4 (n = 4) : $f_4^*(s_4) = \text{Max}\,\{28y_4 - 27x_4 + f_3^*(s_4 + x_4 - y_4)\}$

$= \text{Max}\,\{28y_4 - 27x_4 + 25(s_4 + x_4 - y_4) + 1{,}500\}$

$= \text{Max}\,\{25s_4 - 2x_4 + 3y_4 + 1{,}500\}$

Again comparing the four alternative solutions, the optimal return is obtained at $x_4 = 0$ and $y_4 = s_4$, and

$$f_4^*(s_4) = 28s_4 + 1{,}500$$

However, it is given that, $s_4 = 200$, $x_4 = 0$; $y_4 = 200$. Then,

$s_3 = s_4 - y_4 + x_4$

$= 200 - 200 + 0 = 0;$

$x_3 = 500,\; y_3 = 0$

$s_2 = s_3 - y_3 + x_3$

$= 0 - 0 + 500 = 500;$

$x_2 = 0,\; y_2 = 0$

$s_1 = s_2 - y_2 + x_2$

$= 500 - 0 + 0 = 500;$

$x_1 = 0,\; y_2 = 0$

Thus, the required solution is as follows:

Month	:	1	2	3	4
Purchase	:	0	500	0	0
Sale	:	200	0	0	500

and maximum possible return = 28 (200) + 1,500 = Rs. 7,100.

Example 11:

A company has five salesmen, who have to be allocated to four marketing zones. The return (or profit) from each zone depends upon the number of salesmen working in that zone. The expected returns for different number of salesmen in different zones, as estimated from the past records are given below. Determine the optimal allocation policy.

Number of Salesmen	*Marketing Zones* *1*	*2*	*3*
0	*45*	*30*	*35*
1	*58*	*45*	*45*
2	*70*	*60*	*52*
3	*82*	*70*	*64*
4	*83*	*79*	*72*
5	*101*	*90*	*82*

Solution:

This problem can be treated as a three-stage problem taking each marketing zone as a stage the number of salesmen allocated to a particular zone as a state variable. Let us adopt the following notations:

s = total number of salesmen available

x_j = number of salesmen allocated to marketing zone j (j = 1, 2, 3)

$p_j(x_j)$ = return from zone j when x_j salesmen are allocated, j = 1, 2, 3

Now the given problem can be mathematically stated as:

$$\text{Maximize } Z = p_1(x_1) + p_2(x_2) + p_3(x_3)$$

subject to the constraint

$$x_1 + x_2 + x_3 \leq 5$$

and $\quad x_1, x_2, x_3 \geq 0$ and integers.

To obtain the recurrence relation, let there be s salesmen available to be allocated to all marketing zones. Then $f_j(s)$ represents the optimal allocation of salesmen to marketing zone j and x_j the initial location to marketing zone j. Then the general recursive equation becomes:

$$f_j(s) = \max_{0 \leq x_j \leq s} \{p_j(x_j) + f^*_{j+1}(s - x_j)\};\ j = 1, 2, 3$$

Using backward induction, we start by optimizing the last stage, stage 3 – marketing zone 3. Computations at each stage are shown below:

Stage 2 (j = 3)

s :	0	1	2	3	4	5
$f_3^*(s)$:	35	45	52	64	72	82
x_3^* :	0	1	2	3	4	5

Stage 3 (j = 2) For this stage recursive equation will become

$$f_2(s) = \max_{0 \le x_2 \le s} \{p_2(x_2) + f_3^*(s - x)\}$$

The computations for stages 2 and 1 are shown in Tables 1.16 and 1.17, respectively.

Table 1.16 : Computations for Stage 2

Decision x_2 / *States, s*	$f_2^*(s) = p_2(x_2) + f_2(s - x_2)$ 0	1	2	3	4	5	*Optimal Return*	*Optimal Decision* x_2^*
0	65	–	–	–	–	–	65	0
1	75	80	–	–	–	–	80	1
2	82	90	95	–	–	–	95	2
3	94	97	105	105	–	–	105	2, 3
4	102	109	112	115	114	–	115	3
5	112	117	124	112	124	125	125	5

Stage 1 (j = 1)

Table 1.17 : Computations for Stage 1

Decision x_1 / *States, s*	$f_1^*(s) = p_1(x_1) + f_2^*(s - x_1)$ 0	1	2	3	4	5	*Optimal Return*	*Optimal Decision* x_1^*
5	170	172	175	177	173	166	177	3

In Table 1.17, the maximum return, Rs 177, corresponds to $x_1 = 3$ salesmen, which leaves, s = 5 – 3 = 2 salesmen for two other stages. From Table 1.15, for s = 2, we have $x_2 = 2$. This leaves s= 2 – 2 = 0, *i.e.,* no salesman for stage 1 (marketing zone 3). Hence, the optimal solution so obtained is:

Stage (Marketing zone)	*Value of Decision Variable (Salesmen)*
3 (Zone 3)	0
2 (Zone 2)	2
1 (Zone 1)	3.

Example 12:

The owner of a chain of four grocery stores has purchased six crates of fresh strawberries. The estimated probability distribution of potential sales of the strawberries before spoilage differs among the four stores. The following table gives the estimated total expected profit at each store, when it is allocated various number of crates.

Number of Crates	*Stores*			
	1	*2*	*3*	*4*
0	*0*	*0*	*0*	*0*
1	*4*	*2*	*6*	*2*
2	*6*	*4*	*8*	*3*
3	*7*	*6*	*8*	*4*
4	*7*	*8*	*8*	*4*
5	*7*	*9*	*8*	*4*
6	*7*	*10*	*8*	*4*

For administrative reasons, the owner does not wish to split crates between stores. However, he is willing to distribute zero crates to any of his stores. Find the allocation of six crates to four stores so as to maximize the expected profit.

Solution:

To solve this problem using dynamic programming, we view each store as a stage. We define the following notation.

x_j = number of crates allocated to stage (store) j, (j = 1, 2, 3, 4)

$p_j(x_j)$ = expected profit from allocation of x_j crates to stage (store) j, (j = 1, 2, 3, 4)

Now the problem can be stated mathematically as:

$$\text{Maximize } Z = p_1(x_1) + p_2(x_2) + p_3(x_3) + p_4(x_4)$$

subject to the constraint

$$x_1 + x_2 + x_3 = x_4 = 6 \text{ and } x_1, x_2, x_3, x_4 \geq 0 \text{ and integers.}$$

To obtain the recurrence relation, let there be s crates available to be distributed among stores (stages) and x_j the initial allocation to store j. Then $f_j(s)$ represents the profit associated with the optimal allocation of crates to store (stage) 1 through 4 both inclusive. Thus, the general recurrence equation will become

$$f_j^*(s) = \underset{0 \leq x_j \leq s}{\text{Max}} \{p_j(x_j) + f_{j+1}^*(s - x_j)\};\ j = 1, 2, 3, 4$$

Using backward induction, we start by optimizing the last stage; stage 4, *i.e.*, store 4. The computations at each stage are shown below:

Stage 4 (j = 4)

s	:	0	1	2	3	4	5	6
f_4^* (s)	:	0	2	3	4	4	4	4
x_4^*	:	0	1	2	3	4	5	6

Stage 3 (j = 3)

Table 1.18 : Computations for Stage 3

Decision x_3 / *States, s*	$f_3^*(s) = p_3(x_3) + f_4^*(s - x_3)$ 0	1	2	3	4	5	6	*Optimal Return* $f_3^*(s)$	*Optimal Decision* x_3^*
0	0	–	–	–	–	–	–	0	0
1	2	6	–	–	–	–	–	6	1
2	3	8	8	–	–	–	–	8	1, 2
3	4	9	10	8	–	–	–	10	2
4	4	10	11	10	8	–	–	11	2
5	4	10	12	11	10	8	–	12	2
6	4	10	12	12	11	1	8	12	2, 3

Stage 2 (j = 2)

Table 1.19 : Computations for Stage 2

Decision x_2 / *Stages, s*	$f_2^*(s) = p_2(x_2) + f_3^*(s - x_2)$ 0	1	2	3	4	5	6	*Optimal Return* $f_2^*(s)$	*Optimal Decision* x_2^*
0	0	–	–	–	–	–	–	0	0
1	6	2	–	–	–	–	–	6	1
2	8	8	4	–	–	–	–	8	0, 1
3	10	10	10	6	–	–	–	10	0, 1, 2
4	11	12	12	12	8	–	–	12	1, 2, 3
5	12	13	14	14	14	9	–	14	2, 3, 4
6	12	14	15	16	16	10	15	16	3, 4

Stage 1 (j = 1)

Table 1.20 : Computations for Stage 1

Decision x_1 / Stages, s	0	1	2	3	4	5	6	Optimal Return $f_1^*(s)$	Optimal Decision x_1^*
	$f_1^*(s) = p_1(x_1) + f_2^*(s - x_1)$								
6	16	18	18	17	15	13	7	18	1, 2

In Table 1.20, the maximum return Rs 18 corresponds to the decision, $x_1 = 1$ and $x_1 = 2$ of allocating crates to store 1 (stage 1). When $x_1 = 1$, there remain s = 6 – 1 = 5 crates to be allocated among three other stores. At stage 2, 5 crates (state) corresponds to $x_2 = 2$, 3 or 4. If $x_2 = 2$, then there remain s = 6 – 3 = 3 crates to be allocated to stage 3 and 4. At stage 3, 3 crates corresponds to $x_3 = 2$ and x4 = 1 at stage 4. Hence, one optimal allocation is:

$x_1 = 1$, $x_2 = 2$,
$x_3 = 2$, $x_4 = 1$
$x_1 = 2$, $x_2 = 1$,
$x_3 = 2$, $x_4 = 1$.

Other alternative allocations of crates among stores are as follows to obtain the maximum profit of Rs 18.

Store 1, x_1	:	1	1	1	1	2	2	2	2
Store 2, x_2	:	2	3	3	4	1	2	2	3
Store 3, x_3	:	2	1	2	1	2	1	2	1
Store 4, x_4	:	1	1	0	0	1	1	0	0

Example 13:

Find the minimum value of

$$Z = y_1^2 + y_2^2 + \dots + y_n^2$$

subject to the constraint

$$y_1 \cdot y_2 \cdot \dots y_n = c \; (c \geq 0)$$

and $y_j \geq 0;\; j = 1, 2, \dots, n.$

Solution:

Let $f_n(c)$ be the minimum attainable value of Z when c is divided into n factors.

For n = 1, we have $y_1 = c$ and therefore

$$f_1(c) = \min_{y_1 = c} \{y_1^2\} = c_2$$

For n = 2, let $y_1 = x$ and $y_2 = c/x$. Then

$$f_2(c) = \min_{0 < x \le c} \{x_2 + f_1(c/x)^2\}$$

$$= \min_{0 < x \le c} \{x_2 + f_1(c/x)^2\}$$

Since $f_1(c) = c_2$, we have $f_1(c/x) = (c/x)^2$

For n = 3, let $y_1 = x$, $y_2 \cdot y_3 = c/x$. Then,

$$f_3(c) = \min_{0 < x \le c} \{y_1^2 + y_2^2 + y_3^2\} = \min_{0 < x \le c} \{x_2 + f_2(c/x)\}$$

Proceeding in the same way and using the principle of optimally, the recursive equation will become

$$f_n(c) = \min_{0 < x \le c} \{x_2 + f_{n-1}(c/x)\}; \text{ for all n}$$

Solution of recursive equation : The solution to the recursive equation to get the optimal policy can be obtained with the help of differential calculus. Let $f(x) = x_2 + (c/x)^2$. Then

$$\frac{df}{dx} = 2x - \frac{2c^2}{x^2} = 0$$

which gives $x = (c)^{1/2}$, therefore, $y_1 = (c)^{1/2}$ and $y_2 = c/x = (c)^{1/2}$.

Since the second derivative of f (x) with respect to x is positive, f (x) is minimum. Also

$$f_2(c) = \min_{0 < x \le c} \{x_2 + (c/x)^2\}$$

$$= (c^{1/2})^2 + \left(\frac{c}{c^{1/2}}\right)^2 = 2c$$

Thus $\quad f_2(c/x) = 2(c/x)$

Hence the optimal policy is $(c^{1/2}, c^{1/2})$ and $f_2(c) = 2c$. Again

$$f_3(c) = \min_{0 < x \le c} \{x_2 + f_2(c/x)\} = \min_{0 < x \le c} \{x_2 + 2(c/x)\}$$

$$= (c^{1/3})^2 + 2\left(\frac{c}{c^{1/3}}\right) = 3c^{2/3}$$

since minimum of $f(x) = x_2 + 2c/x$ was obtained at $x = c^{1/3}$. Hence the optimal policy is: $\{c^{1/3}, c^{1/3}, c^{1/3}\}$ and $f_3(c) = 3c^{1/3}$.

Continuing in this manner, the optimal policy for an n stage problem will be

$$\{c^{1/n}, c^{1/n}, \ldots, c^{1/n}\} \text{ and } f_n(c) = nc^{2/n}$$

Example 14:

Use dynamic programming to solve the following linear programming problem.

Maximize $Z = 3x_1 = 5x_2$

subject to the constraints

$$x_1 \le 4$$

$$x_2 \le 6$$

$$3x_1 + 2x_2 \le 18$$

and $x_1, x_2 \ge 0$

Solution:

This linear programming problem can be considered as a two-stage, three-stage problem because there are two decision variables and three constraints with available resources.

The optimal value of $f_1(b_1, b_2, b_3)$ at the first stage is given by

$$f_1(b_1, b_2, b_3) = \underset{0 \le x_1 \le b}{\text{Max}} \{3x_1\}$$

where $b_1 = 4$, $b_2 = 6$ and $b_3 = 18$.

The feasible value of x_1 is a non-negative value which satisfies all the given constraints $x_1 \le b_1$ (= 4), $3x_1 \le b_3$ (= 18). Thus, the maximum value of b that x_1 can assume is, b = Min (4, 18/3) = 4. Therefore,

$$f_1(4, 6, 18) = \underset{0 \le x_1 \le 4}{\text{Max}} \{3x_1\}$$

$$= 3 \text{ Min} \left\{4, 6 - \frac{2}{4}.x_2\right\}$$

and $$x_1^* = \text{Min} \left\{4, 6 - \frac{2}{3}.x_2\right\}$$

The recursive relation for optimization of this two-stage problem is

$$f_2(b_1, b_2, b_3) = \underset{0 \le x_2 \le b}{\text{Max}} \{5x_2 + f_1^*(b_1, b_2 - x_2, b_3 - 2x_2\}$$

where the maximization of x_2 satisfying the conditions of $x_2 \le b_2$ (= 6) and $2x_3$ (= 18) is the minimum of b = min (6, 9) = 6. Therefore, the recurrence relationship can be expressed as:

$$f_2(4, 6, 18) = \underset{0 \le x_2 \le 6}{\text{Max}} \{5x_2 = f_1^*(4, 6 - x_2, 18 - 2x_2)\}$$

$$= \underset{0 \le x_2 \le 6}{\text{Max}} \left\{5x_2 + 3\text{Min}\left(4, 6\frac{2}{3}x_2\right)\right\}$$

Since $\text{Min}\left(4, 6 - \frac{2}{3}.x_2\right)$

$$= \begin{cases} 4; & 0 \le x_2 \le 3 \\ 6 - \frac{2}{3}x_2; & 3 \le x_2 \le 6 \end{cases} = \begin{cases} 4; & 0 \le x_2 \le 3 \\ 6 - \frac{2}{3}x_2; & 3 \le x_2 \le 6 \end{cases}$$

we get $\text{Max}\left\{5x_2 + 3\text{Min}\left(4, 6 - \frac{2}{3}x_2\right)\right\} = \begin{cases} 5x_2 + 12; & 0 \le x_2 \le 3 \\ 18 + 3x_2; & 3 \le x_2 \le 6 \end{cases}$

Now the maximum value of $5x_2 + 12 = 27$ at $x_2 = 3$ and maximum value of $18 + 3x_2 = 36$ at $x_2 = 6$. Therefore, the optimal value of $f_2^* = (4, 6, 18) = 36$ is obtained at $x_2 = 6$. Since

$$x_1^* = \text{Min}\left\{4, 6 - \frac{2}{3}x_2\right\} = \text{Min}\left\{4, 6 - \frac{2}{3} \times 6\right\} = 2$$

The optimum solution to the given LP problem is; $x_1 = 2$, $x_2 = 6$ and Max Z = 36.

Example 15:

Solve the following LP problem by dynamic programming approach

Maximize $Z = 8x_1 + 7x_2$

subject to the constraints

$$2x_1 + x_2 \le 8$$

$$5x_2 + 2x_2 \le 15$$

and $x_1, x_2 \ge 0$

Solution:

This LP problem can be considered as a two-stage, two-state problem because there are two decision variables and two constraints. Starting with the second stage backward, the procedure is as follows:

The optimal value of f_2 (b_1, b_2) at the second stage is given by

$$f_2\ (b_1, b_2) = \underset{0 \le x_2 \le b}{\text{Max}} \{7x_2\}$$

where $b_1 = 8$, and $b_2 = 15$. The feasible value of x_2 is a non-negative value which satisfies all the given constraints $x_2 \le b_1$ (= 8) and $2x_2 \le b_2$ (= 15). Thus the maximum value of b that x_2 can assume is: b = min (8, 15/2) = 15/2. Therefore

$$f_2\ (b_1, b_2) = \underset{0 \le x_2 \le b}{\text{Max}} \{7x_2\}$$

$$= 7 \text{ Min } \{8 - 2x_1, (15/2) - (5/2)\ x_1\}$$

and $$x_2^* = \text{Min } \{8 - x_4, 7.5 - 2.5\ x_1\}$$

Proceeding backwards to stage 1 (j = 1), the recursive relation for optimization can be expressed as:

$$f_1\ (b_1, b_2) = \underset{0 \le x_1 \le b}{\text{Max}} \{8x_1 + f_2^*\ (b_1 - 2x_1, b_2 - (5/2)x_1)\}$$

$$= \underset{0 \le x_1 \le 3}{\text{Max}} \{8x_1 + 7 \text{ Min } (8 - 2x_1; (15/2) - (5/2)x_1)\}$$

where maximization of variable x_1 satisfying the conditions: $2x_1 \le b$ (= 8) and $5x_1 \le b_2$ (= 15) is the minimum of b = min (8/2, 15/5) = 3. Since the minimum (*i.e.*, zero) of $(8 - 2x_1, 15/2 - 5x_1/2)$ is obtained at $x_1 = 3$ for $0 \le x_1 \le 3$, we get

$$f_1^*\ (b_1, b_2) = \text{Max } [8x_1 + 7 \text{ Min } \{8 - 2x_1; (15/2) - (5x_1/2)\}]$$

$$= \text{Miax } \{8 \times 3 + 7 \times 0\} = 24, \text{ at } x_1 = 3$$

and $$x_2^* = \text{Min } \{8 = 2x_1, 7.5 - 2.5x_1\}$$

$$= \text{Min } \{8 - 6, 7.5 - 2.5\ (3)\} = 0$$

Hence, the optimum solution to the given LP problem is: $x_1 = 3$, $x_2 = 0$ and Max Z = 94.

EXERCISES

1. State Bellman's "principle of optimality" and explain by an illustrative example how it can be used to solve a multi-stage decision problem.
2. Discuss dynamic programming with suitable examples.
3. Define the following dynamic programming terms:

   (a) Stage (b) State variable

(c) Decision variable (d) Immediate return

(e) Optimal return (f) State transformation function.

4. What is the dynamic recursive relation? Describe the general process of backward recursion.

5. Why is it frequently desirable to solve a problem with a number of decision variables by dividing it into a series of sub-problems?

6. (a) Explain the concept of dynamic programming and the relation between dynamic and linear programming approach.

   (b) Illustrate with example the method for solving a linear programming problem by the dynamic programming approach.

7. (a) Discuss dynamic programming with suitable examples.

   (b) Define a standard warehouse problem. Outline a procedure to solve it.

8. Discuss briefly:

   (a) The general similarities between dynamic programming and linear programming.

   (b) How does dynamic programming differ conceptually from linear programming?

9. Use dynamic programming to find the value of

$$\text{Max } Z = y_1 \cdot y_2 \cdot y_3$$

subject to the constraint

$$y_1 + y_2 + y_3 = 5$$

and $$y_1, y_2, y_3 \geq 5$$

10. Show how the functional equation technique of dynamic programming can be used to determine the shortest route when it is constrained to pass through a set of specified nodes which is a definite subset of the set of nodes of a given network.

11. State Bellman' principle of optimality and apply it to solve the following problem

$$\text{Max } Z = x_1 \,.\, x_2 \ldots x_n$$

subject to the constraint

$$x_1 + x_2 + \ldots, + x_n = c$$

and $$x_1, x_2, + \ldots, + x_n \geq 0$$

12. A government space project is conducting research on a certain engineering problem that must be solved before a man can fly to moon safely.

    These research teams are currently trying three different approaches for solving the problems. An estimate has been made that, under present circumstances, the probability that the respective teams; say A, B and C will not succeed are 0.40, 0.60 and 0.80, .respectively. Thus, the current probability that all three teams will fail is (0.40) (0 60) (0.80) = 0.192. Since the objective is to minimize this probability, the decision has been made to assign two or more top scientists among the three teams in order to lower the probability of failure as much as possible.

    The following table gives the estimated probability that the respective teams will fail when 0, 1 or 2 additional scientists are added to that team.

Number of new scienntists	Team A	Team B	Team C
0	0.40	0.60	0.80
1	0.20	0.40	0.50
2	0.15	0.20	0.30

    How should the additional scientists be allocated to the team?

13. A truck can carry a total of 10 tonnes of product. Three types of products are available for shipment. Their weights and values are tabulated. Assuming that at least one of each type must be shipped, determine the loading which will maximize the total value.

Product Type	*Value (Rs)*	*Weight (Rs)*
A	20	1
B	50	2
C	60	3

14. Consider the problem of designing an electronic device consisting of three main components. The components are arranged in series so that the failure of one of them will result in the failure of the whole device. Therefore, it is decided that the reliability (probability of failure) of the device can be increased by installing parallel units on each component. Each component may be installed in at the most three parallel units. The total capital (in thousand Rs) available for the device is 10. The following data is available:

Number of Parallel Units, m_i	*Components*					
	1		*2*		*3*	
	r_1	c_1	r_2	c_2	r_3	c_3
1	0.50	2	0.70	3	0.60	1
2	0.70	4	0.80	5	0.80	2
3	0.90	5	0.90	6	0.90	3

where r_i, c_i (i = 1, 2, 3) is the reliability and the cost of the ith component, respectively. Determine the number of parallel units which will maximize the total reliability of the system without exceeding the given capital.

15. Use the principle of optimality to find the maximum value of

    Max $Z = b_1x_1 + b_2x_2 + ... + b_nx_n$

    subject to the constrain = $x_1 + x_2 + ... + x_n = c$

    and $x_1, x_2, ..., x_n \geq 0$

16. A student has to take an examination in three courses x, y and z. He has three days available for study. He feels that it would be better to devote a whole day to study of the same course, so that he may study a course for one day, two days or three days or not at all. His estimates of grades he may get according to days of study he puts in are as follows:

Study Days	*Course*		
	x	*y*	*z*
0	1	2	1
1	2	2	2
2	2	4	4
3	4	5	4

17. A chairman of a certain political party is making plans for his election to the Parliament. He has engaged the services of six volunteer workers and wishes to assign them four districts in such a way as to maximize their effectiveness. He feels that it would be inefficient to assign a worker to more than one district but he is also willing to assign no worker to any one of the districts judging by what the workers can accomplish in other districts.

    The following table gives the estimated increase in the number of votes in favour of the party's candidate if it were allocated various number of workers;

Number of Workers	*Districts* 1	2	3
0	0	0	0
1	25	20	33
2	42	38	43
3	55	54	47
4	63	65	50
5	69	73	52
6	74	80	53

How many of the workers should be assigned to each of the three districts in order to maximize the total number of votes in his favour?

18. We have a machine that deteriorates with age and so we need to formulate a replacement policy. We have to own such a machine during each of the next years. The operating cost c(i) of a machine i years old at the beginning of the year, trade in value t(i) received when such a machine is traded for a new machine at the start of the year and s(i), the salvage value received for a machine that have just turned age i at the end of 5 years are given below:

i	:	0	1	2	3	4	5	6
c(i)	.	10	13	20	40	70	100	100
t(i)	:	–	32	21	11	5	0	0
s(i)	:	–	25	17	8	0	0	0

If a new machine costs Rs 50 and we have now a machine which is two years old, what is the optimum policy of replacement? Solve the problem by dynamic programming.

19. The following table provides the monthly demand for an item of a company during the winter season. Virtually all direct labour is casual and temporary so that any number of items up to the plant capacity of 400 items may be produced each month. The direct labour cost is Rs 40 per item.

Month	:	November	December	January	February
Demand	:	100	200	300	400

The plant is completely shut down during a month when there is no production, and supervisory personnel are given an unpaid holiday to save a further Rs 2,000 in monthly payroll costs. Demands may be filled either from current production or from inventory. Each item held in inventory at a beginning of the month costs the company Rs 5. The production lot size may be 0, 100, 200, 300 or 400 items. Determine the monthly production levels in such a way that total cost is minimized,

20. A company is planning its manufacturing operations for the next five months. The following demands apply:

Month	:	January	February	March	April	May
Demand	:	200	300	300	200	400

Each item held in inventory from one month to the next incurs a Rs 4 carrying charge. There are 200 unsold items remaining at the end of December. The company is to redesign its production for June, so that a no ending inventory is desired. A maximum of 400 units may be manufactured in any given month. The total production costs are:

Size of production	:	0	100	200	300	400
Cost (Rs)	:	0	1,000	1,300	1,450	1.525

Determine the minimum cost production plan.

21. A shoe store sells rubber shoes for a particular season which lasts from December 1 through February 29. The sales division has forecast the following demands for the next year.

Month	:	December	January	February
Demand	:	30	40	30

All shoes sold by the store are purchased from outside sources. The following information is known about this particular shoe.

(i) The unit purchasing cost is Rs 100 per pair; however, the supplier will only sell in lots of 10, 20, 30, or 40 pairs. Any orders for more than 50 or less than 10 will not be accepted.

(ii) The following quantity discounts apply on lot size orders:

Lot size	:	10	20	30	40
Discount (per cent)	:	5	5	10	20

(iii) For each order placed, the store incurs a fixed cost of Rs 20. In addition, the supplier charges an average amount of Rs 50 per order to cover transportation costs, insurance, packaging and so on, irrespective of the amount ordered.

(iv) Due to large in-process inventories, the store will carry no more than 40 pairs of shoes in inventory at the end of any one month. Carrying charges are Rs 5 per pair per month, based on the end of month inventory. It is desirable to have both incoming and outgoing seasonal inventory at zero.

Find an ordering policy which will minimize total seasonal costs.

22. A pharmaceutical company has ten medical representatives working in three sales areas. The profitability for each representative in three sales areas is as follows:

No. of Representatives		:	***0***	***1***	***2***	***3***	***4***	***5***	***6***	***7***	***8***
Profitability	Area 1	:	15	22	30	38	45	48	54	60	65
(Rs '000)	Area 2	:	26	35	40	46	55	62	70	76	83
	Area 3	:	30	38	44	50	60	65	72	80	85

Determine the optimum allocation of medical representatives in order to maximize the profits. What will be the optimum allocation to the number of representatives available at present is only six?

23. A company has three media A, B and C available for advertising its product. The data collected over the past years about the relationship between the sales and frequency of advertisement in the different media is as follows:

Frequency/Month	***Estimated***	***Sale (units) per Month***	
	A	***B***	***C***
1	125	180	300
2	225	290	350
3	260	340	450
4	300	370	500

The cost of advertisement is Rs 5,000 in medium A, Rs 10,000 in medium B and Rs 20,000 in medium C. The total budget allocated for advertising the product is Rs 40,000. Determine the optimal combination of advertising media and frequency.

24. Neema-Chem manufacturers use cupric chloride as the basic material for the production of copper complex. For a smooth functioning of its production schedules, the company may have enough stock every month of its basic material. The purchase price p_n and the demand d_n forecast for the next 6 months by the management is given below:

Month (n)	:	1	2	3	4	5	6
Purchase price (p_n)	:	11	18	13	17	20	10
Demand (d_n)	:	8	5	3	2	7	4

Due to limited space, the warehouse cannot carry more than 9 units of the basic material. The basic material is bought at the beginning of each month. When the initial stock is 2 units and the final stock is required to be zero, the company wishes to find an ordering policy for the next 6 months so as to minimize the total purchase cost.

25. An enterprising researcher believes that he has developed a system for winning a popular Las Vegas game. His colleagues do not believe that this is possible, so they have made a large bet with him. They bet that, starting with three chips, he will not have at least five chips after three plays of the game. Each play of the game involves betting any desired number of available chips and then either winning or losing this number of chips. He believes that his system will give him a probability of 2/3 of winning a given play of the game. Find his optimal policy regarding how many chips to bet, if any, at each of the three plays of the game in order to maximize the probability of winning his bet with his colleagues.

26. The World Health Council is devoted to improving health care in the underdeveloped countries of the world. It now has five medical teams available to allocate among three such countries to improve their medical care, health education and training programmes. Therefore, the council needs to determine how many teams (if any) to allocate to each of these countries to maximize the total effectiveness of the five teams. The measure of effectiveness being used is additional man-years of life. (For a particular country, this measure equals the country's increased life expectancy in years times its population.) The following table gives the estimated additional man-years of life (in multiple of 1,000) for each country for each possible allocation of medical teams.

Number of Medical Teams	*Thousands of Additional Man-years of Life*		
	Country 1	*Country 2*	*Country 3*
0	0	0	0
1	45	20	50
2	70	45	70
3	90	75	80
4	105	110	100
5	120	150	130

Determine how many teams are to be allocated to each country for maximum effectiveness. Also form the recursive equation.

27. A man is engaged in buying and selling identical items, each of which requires considerable storage space. The buying and selling prices are indicated in the table below. He operates from a warehouse which has a capacity of 500 items. He can order on the 15th of each month, for delivery on the first day of the following month. During a month, he can also sell any amount up to his total stock on hand.

	January	*February*	*March*
Cost Price (Rs)	150	155	165
Sales Price (Rs)	165	165	185

If he starts the year with 200 items in stock, how much should he plan to purchase and sell each month, in order to maximize his profit for the first quarter of the year?

28. An investor has Rs. 6,000 to invest. This amount can be invested in any of the three ventures available to him. But, he must invest in units of Rs 1,000. The potential return from investment in any one venture depends upon the amount invested according to the following table (all figures in thousands).

Amount Invested	*Return from Venture*		
	A	*B*	*C*
0	0	0	0
1	0.5	1.5	1.2
2	1.0	2.0	2.4
3	3.0	2.2	2.5
4	3.1	2.3	2.6
5	3.2	2.4	2.7
6	3.3	2.5	2.8

Formulate the above problem as a dynamic programming problem and find the optimum investment policy.

2

Convex Sets

INTRODUCTION

Definition (Vectors) : Any ordered n-tuple of numbers is called *an n-vector*. By an ordered n-tuple we mean a set consisting of numbers in which the place of each number is fixed. If $x_1, x_2, \ldots x_n$ be any n numbers, then the ordered n-tuple $X = (x_1, x_2, \ldots, x_n)$ is called an n-vector. The ordered triad (x_1, x_2, x_3) is called a 3-vector. Similarly (1, 0, 1, – 1) and (1, 8, – 5, 7) are 4-vectors. The n numbers $x_1, x_2, \ldots$,mn are called components of the n-vector $X = (x_1, x_2, \ldots, x_n)$. A vector may be written either as a row vector or as a column vector. If A be a matrix of the type $m \times n$, then each row of A will be an n-vector and each column of A will be an m-vector. A vector whose components are all zero is called a zero vector and will be denoted by O.

If k be any number and X be any vector, then relative to the vector X. k is called a scalar.

Equality of Two Vectors

Two n-vectors X and Y where $X = (x_1, x_2, \ldots x_n)$ and $Y = (y_1, y_2, \ldots y_n)$ are said to be *equal* if and only if their corresponding components are equal *i.e.,* if $x_i = y_i$, for all $i = 1, 2, \ldots, n$.

For example if,

$$X = (1, 4, 7) \text{ and } Y = (1, 4, 7),$$

then $\quad X = Y.$

Properties of Addition and Scalar Multiplication of Vectors

If X, Y, Z be any three n-vectors and p, q be any two numbers, then obviously

(i) $X + Y = Y + X$.

(ii) $X + (Y + Z) = (X + Y) + Z$.

(iii) $p (X + Y) = pX + pY$.

(iv) $(p + q) X = pX - qX$.

(v) $p (qX) = (pq) X$.

Multiplication of a Vector by a Scalar (Number)

If k be any number and

$$X = (x_1, x_2, \ldots x_n),$$

then by definition

$$k\,X = (kx_1, kx_2, \ldots, kx_n).$$

The vector k X is called the scalar multiple of the vector X by the scalar k.

Algebra of Vectors

Since an n-vector is nothing but a row matrix or a column matrix. Therefore we can develop an algebra of vectors in the same manner as the algebra of matrices.

DEFINITIONS

1. **An ∈-neighbourhood.** *An neighbourhood about the point a is defined as the set of points lying inside the hypersphere with centre at a and radius e > 0.*

   i.e., the Î-neighbourhood about the point a is the set of points

   $X = \{x: |x - a| < \varepsilon\}.$

2. **An Interior Point.** *A point a is an interior point of the set S if there exists an ε-neighbourhood about a which contains only points of the set S.*

   An interior point of S must be an element of S.

3. **Boundary Point.** *A point a is a boundary point of the set S if every ε-neighbourhood about a (e > 0 may be, however, small) contains points which are in the set and the point which are not in the set.*

   A boundary point of S does not have to be an element of S.

4. **Point Sets.** *Point sets are sets whose elements are points or vectors in E^n or R^n (n-dimensional Euclidean space).*

   For example,

   (i) A linear equation in two variables x_1, x_2, *i.e.*, $a_1 x_1 + a_2 x_2 = b$ represents a line in two dimensions. This line may be considered as a set of those points (x_1, x_2) which satisfy $a_1 x_1 + a_2 x_2 = b$. This set of points can be written as

   $S_1 = \{(x_1, x_2): a_1 x_1 + a_2 x_2 = b\}.$

   (ii) Consider the set of points lying inside a circle of unit radius with centre at the origin, in two dimensional space (E^2). Obviously the points (x_1, x_2) of this set satisfy the inequality $x_1^2 + x_2^2 < 1$.

This set of points can be written as

$$S_2 = \{(x_1, x_2): x_1^2 + x_2^2 < 1\}.$$

These sets may contain either a finite or infinite number of elements. Usually, however, they will contain an infinite number of elements. Further, we shall always assume that there is at least one element in a set unless otherwise stated.

5. **Hypersphere.** *A hypersphere in En with centre at a and radius ε > 0 is defined to be the set of points*

$$X = \{x: |x - a| = \varepsilon\}$$

i.e., the equation of a hypersphere in E^n is

$$(x_1 - a_1)^2 + (x_2 - a_2)^2 + ... + (x_n - a_n)^2 = \varepsilon^2$$

where $a = (a_1, a_2, ... a_n)$,

$x = (x_1, x_2, ... x_n)$,

which represents a circle in E^2 and sphere in E^3.

The set of points inside the hypersphere in E^3.

The set of points inside the hypersphere is the set

$$X = \{x: |x - a| < \varepsilon \}.$$

CONVEX SETS AND THEIR PROPERTIES

Definition *(Convex Set). A subset SCR^n is said to be convex is convex self all their properties for each pair of points x, y in S, the line segment [x : y] is contained in S.*

In symbols, a subset SCR^n is convex iff

$$x, y \varepsilon S \Rightarrow [x : y] CS.$$

1. **A Closed Set.** *A set S is said to be a closed set if it contains all its boundary points.*
2. **Lines, In E^n.** *The line through the two points x_1 and x_2, $x_1 \neq x_2$ is defined to be the set of points*

$$X = \{x:x = \lambda x_1 + (1 - \lambda) x_2, \text{ for all real } \lambda\}.$$

3. **An Open Set.** *A set S is said to be an open set if it contains only the interior points.*
4. **Line Segments.** *In E^n, the line segment joining two points x1 x_2 is defined to be the set of points*

$$X = \{x : x = \lambda\, x_1 + (1 - \lambda)\, x_2,\quad 0 \le y \le 1\}.$$

Note that the restriction $0 \le y \le 1$ restricts the point x to lie within the in joining the points x_1 and x_2. Line segment of x_1, x_2 is also denoted by $[x_1 : x_2]$.

Note that the restriction $0 \le l \le 1$ restricts the point x to lie within the line joining the points x_1 and x_2. Line segment of x_1, x_2 is also denoted by $[x_1 : x_2]$.

5. **Hyperplane.** *A* ***hyperplane*** *is defined as the set of points satisfying*

$$c_1 x_1 + c_2 x_2 + \ldots + c_n x_n = z \text{ (not all } c_i = 0)$$

or $\quad c\, x = z$

for prescribed values of $c_1, c_2, \ldots, c_n$ *and* z.

For optimum value of z this hyperplane is called optimal hyperplane.

The vector c is called a *vector normal to the hyperplane* and $\pm \dfrac{c}{|c|}$ are called unit normal.

It can be easily seen that hyperplanes are closed sets.

A hyperplane divides the whole space E^n into three mutually disjoint sets given by

$$X_1 = \{x : c\, x > z\}$$
$$X_2 = \{x : c\, x = z\}$$
$$X_3 = \{x : c\, x < z\}.$$

The sets X_1 and X_2 are called open half spaces.

The sets $\{x : c\, x \le z\}$ and $\{x : c\, x \ge z\}$ are called closed half spaces.

Note: The objective function of a L.P.P. represents a hyperplane and each constraint (sign $\le$, $\ge$) is a closed half space produced by the hyperplane given by the constraint by taking (=) sign in place of $\le$ and $\ge$.

Parallel Hyperplanes. *Two hyperplanes* $c_1 x = z_1$ *and* $c_2 x = z_2$ *are said to be parallel if they have the same unit normals i.e., if* $c_1 = \lambda\, c_2$ *for some* λ, λ *being non-zero.*

CONVEX COMBINATION

A convex combination of a finite number of points $x_1, x_2, \ldots, x_n$ is defined as a point

$$x = \lambda_1 x_1 + \lambda_2 x_2 + \ldots + \lambda_n x_n,$$

where λ_i *is real and* ≥ 0, $\forall\, i$ *and* $\sum_{i=1}^{n} \lambda_i = 1$.

The convex linear combination of two points x_1 and x_2 is given by

$$x = \lambda_1 x_1 + \lambda_2 x_2, \text{ s.t. } \lambda_1, \lambda_2 \geq 0, \lambda_1 + \lambda_2 = 1.$$

It can also be written as

$$x = \lambda x_1 + (1 - \lambda) x_2, 0 \leq l \leq 1.$$

This shows that the line segment of the two points x_1 and x_2 is nothing but the set of all possible convex combinations of the two points x_1 and x_2.

CONVEX SET

A set of points is said to be convex if for any two points in the set, the line segment joining these two points is also in the set. In other words a set is convex if the convex combination of any two points in the set, is also in the set.

If can be easily seen that the convex combination of any number of points in the convex set also belongs to the set.

By convention a set of one point is always convex.

CONVEX FUNCTION

A function f (x) is said to be strictly convex at x if for any two other distinct points x_1 and x_2

$$f\{\lambda x_1 + (1 - \lambda) x_2\} < l\, f(x_1) + (1 - \lambda) f(x_2), \text{ where } 0 < \lambda < 1.$$

On the other hand, a function f (x) is strictly concave if – f (x) is strictly convex.

The following illustration should help to observe the relation of a set A to < A >.

EXTREME POINT OF A CONVEX SET

A point x in a convex set C is called on *extreme point* if x cannot be expressed as a *convex combination* of any two *distinct points* x_1 *and* x_2 *in C.*

In other words, a point x in a convex set C is an extreme point of C if it does not lie on the line segment of any two points, different from x in the set.

Mathematically, a point x is an extreme point of a convex set if there do not exist other points x_1, x_2 ($x_1 \neq x_2$) in the set such that

$$x = \lambda x_1 + (1 - \lambda) x_2, 0 > \lambda < 1.$$

For example : The set $C = \{(x_1, x_2) : x_1^2 + x_2^2 \leq 1\}$ is convex. Every point on the circumference is an extreme point. Thus a convex set may also have infinite number of extreme points.

The polygons which are convex sets have the extreme points as their vertices.

Obviously, an extreme point is a boundary point of the set.

It is important to note that all boundary points of a convex set are not necessarily extreme points.

A point of C which is not an extreme point, is referred as an internal point of C.

CONVEX HULL

The convex hull C(X) of any given set of points X is the set of all convex combinations of sets of points from X.

In other words, the intersection of all convex sets, containing $X \subset E^n$, is called the convex hull of X and is denoted by $< X >$. Thus, the convex hull of a set $X \subset E^n$, is the smallest convex set containing X.

LINEAR DEPENDENCE AND LINEAR INDEPENDENCE OF VECTORS

Linearly Dependent Det of Vectors

Definition : *A set r n-vectors $X_1, X_2, \dots X_r$ is said to be linearly dependent if there exist r scalars (numbers) $k_1, k_2, \dots k_r$, not all zero, such that*

$$k_1 X_1 + k_2 X_2 + \dots + k_r X_r = O,$$

where, O, denotes the n-vector whose components are all zero.

Linearly Independent Set of Vectors

Definition : *A set r n-vectors $X_1, X_2, \dots X_r$ is said to be linearly independent if every relation of the type*

$$k_1 X_1 + k_2 X_2 + \dots + d_r X_r = O$$

implies $k_1 = k_2 = k_3 = \dots = k_r = 0.$

A VECTOR AS A LINEAR COMBINATION OF VECTORS

Definition : *A vector X which can be expressed in the form*

$$X = k_1 X_1 + k_2 X_2 + \dots + k_r X_r,$$

is said to be a linear combination of the set of vectors

$$X_1, X_2, \dots, X_r.$$

Here $k_1, k_2, \dots, k_r$ are any numbers.

The following two results are quite obvious:

(i) *If a set of vector is linearly dependent, then at least one member of the set can be expressed as a linear combination of the remaining members.*

(ii) *If a set of vectors is linearly independent then no member of the set can be expressed as a linear combination of the remaining members.*

Definition 1: *(Equality of two n-vectors).* Two n-vectors $a = [a_1, a_2, \ldots, a_n]$ and $b = [b_1, b_2, \ldots b_n]$ are said to equal if and only if $a_i = b_i$ for all $i = 1, 2, \ldots, n$ note that if $a = b$, then $b = a$.

Definition 2: *(Sum of two vectors).* The sum of two n-vectors $a = [a_1, a_2, \ldots a_n]$ and $b = [b_1, b_2, \ldots, b_n]$, written as $a + b$, is the n-vector

$$a + b = [a_1 + b_1, a_2 + b_2, \ldots, a_n + b_n].$$

It is easy to verity the following properties of vector addition:

(i) for all $a, b \in R^n$,

$$a + b = b + a$$

(ii) for all $a, b\ c \in R^n$,

$$a + (b + c) = (a + b) + c$$

(iii) There exist the zero vector $0 = [0, 0, \ldots, 0]$ such that for all $a \in R^n$,

$$a + 0 = 0 + a = a.$$

Definition 3: *(n-vectors) An N-vector a is an ordered n-taple of real numbers, written in the form of a column,*

$$a = \begin{bmatrix} a_1 \\ a_2 \\ \cdots \\ \cdots \\ \cdots \\ a_n \end{bmatrix} \text{ or } [a_1, a_2, \ldots, an]$$

The numbers a_i ($i = 1, 2, \ldots n$) are called its components.

The set of all n-vectors is denoted by the symbol R^n.

The notation $[a_1, a_2, \ldots a_n]$ for an n-vector indicates and n-type written in the form of a column. It should not be confused with the notation $(a_1, a_2, \ldots, a_n)$ for a paint.

However, there is a connection, in that the n-vector a, that corresponds to the point

$A = (a_1, a_2, \ldots, a_n)$, is the vector.

$$a = [a_1, a_2, \ldots, a_n].$$

Definition 4: *Unit vector. A vector n is acid to be unit sector if* $|n| = 1$

Definition 5: *Angle between two vectors. The angle* θ *between two vectors is defined*

$$\cos\theta = \frac{a.b}{|a|.|b|} \quad 0 \le \theta \le \Pi.$$

Definition 6: *(Multiplication of a vector by a scalar).* The Scalar Multiple of an n-vector, $a = [a_1, a_2, \ldots, a_n]$ by a scalar λ, written as λ a, is defined to be the n-vector

$$\lambda\, a = [\lambda a_1, \lambda a_2, \ldots, \lambda a_n]$$

Thus $\lambda\, a = -a$, if $\lambda = -1$

and $\lambda\, a = 0$ if $\lambda = 0$.

The operation of finding a scalar multiple satisfies the following properties.

(i) For all a, b $\in R^n$, and scalar c,

$c\,(a + b) = c_a + c_b$

(ii) For any vector a $\in R^n$, and scalar c_1, c_2

$(c_1 + c_2)\, a = c_{1a} + c_{2a}$

(iii) For any vector a $\in R^n$, and scalars c_1, c_2

$c_2\,(c_1\, a) = (c_2\, c_1)\, a = c_1\,(c_2\, a)$

(iv) For any vector a $\in R^n$,

$1\, a = a$

We omit the verification, which is straight forward.

Definition 7: *(Inner product or scalar product).* Given two n-vectors, $a = [a_1, a_2, \ldots, a_n]$ and $b = [b_1, b_2, \ldots b_n]$ the *Inner (Scalar) Product* of a and b, to be denoted by a. b is the scalar

$$a.\, b = a_1\, b_1 + \ldots + a_n\, b_n$$

$$\text{or } a.\, a = \sum_{i=1}^{\lambda} a_i b_i$$

It is easy to verity the following:

(i) a. b = b. a

(ii) (a + b).c = a.c + b.c

(iii) a.l (b) = λ. (a. b) = (λa.). b

Definition 8: *(The length of a vector).* The length (or the norm) or a vector a $\in R^n$ to be denoted by $|a|$ is defined as $|a| = \sqrt{a.a}$

$$|a| = \sqrt{a_1^2 + a_2^2 + \ldots a_n^2}.$$

ANALYTICAL METHOD

The L.P. problems having more than two variables cannot be solved by Graphical method. In such cases, the analytic solution (trial and error method) may seem to be useful. For solving a L.P.P. Simplex method is the most powerful technique (discussed in the next chapter).

Extreme point theorem states that an optimal solution to a L.P.P. occurs at one of the vertices of the feasible region.

Since the vertices of the feasible region correspond to the basic feasible solutions., the objective function is optimal at least at one of the basic solutions. Some of the vertices may be infeasible which are dropped from consideration.

Consider the following L.P.P.

Maximize $Z = -0.10x_1 + 0.50x_2$

subject to the conditions $2x_1 + 5x_2 \leq 80$

$x_1 + x_2 \leq 20$

$x_1, x_2 \geq 0.$

Introducing the slack variables, the given L.P.P. reduces to the following form:

Maximize $Z = -0.10x_1 + 0.50x_2 + 0.x_3 + 0.x_4$

subject to $2x_1 + 5x_2 + x_3 = 80$

$x_1 + x_2 + x_4 = 20$ ...(1)

$x_1, x_2, x_3, x_4 \geq 0.$

We know that for a system of m equations in n unknowns (n > m), a solution is called a basic solution if in it at least (n – m) variables have the value zero. This gives a vertex.

Taking (n – m) variables equal to zero at a time and solving the resulting system of equations we get a basic solution.

Here n = 4 (the number of variables) and m = 2 (the number of equations). Out of 4 variables 2 can be taken in $^4C_2 = 6$ number of ways. Hence at the most there will be 6 basic solutions which can be obtained as follows:

1. Taking $x_2 = x_4 = 0$, system (1) gives $x_1 = 20, x_3 = 40$.
2. Taking $x_1 = x_4 = 0$, system (1) gives $x_2 = 20, x_3 = -20$.

3. Taking $x_1 = x_3 = 0$, system (1) gives $x_2 = 16, x_4 = 4$.
4. Taking $x_1 = x_2 = 0$, system (1) gives $x_3 = 80, x_4 = 20$.
5. Taking $x_3 = x_4 = 0$, system (1) gives $x_1 = 20/3, x_2 = 40/3$.
6. Taking $x_2 = x_3 = 0$, system (1) gives $x_1 = 40, x_4 = -20$.

Since basic solutions 3 and 5 so not satisfy the non-negative restriction of the variables so these do not give the B.F. solutions and so are dropped from the consideration. Thus the B.F. solutions of the given L.P.P. are given by 1, 2, 4 and 6. For these B.F. solutions the values of the objective function Z are respectively

$$Z_1 = 0, Z_2 = 6, Z_4 = -2 \text{ and } Z_6 = 8.$$

Since maximum $Z = Z_6 = 8$, so the optimal solution to the given L.P.P. is given by $x_1 = 0$, $x_2 = 16$ and max. $Z = 8$.

The same solution is obtained earlier also by graphical method.

Here it is important to note that the B.F. solutions 1, 2, 4 and 6 exactly coincide with the corner points of the feasible region and one of these corner points gives the optimal solution.

Drawbacks of Analytical Method

(i) For large values of m and n, it is extremely difficult and time consuming to solve various sets simultaneous equations.

(ii) Some of the sets give infeasible also which are dropped from consideration.

There is no technique to detect all such sets and not to solve them at all.

(iii) We observe that the value of Z changes from 0 to 6 to – 2 to 8 *i.e.*, there are ups and downs.

LINEAR COMBINATION OF VECTORS

Given two vectors a and b, we can always perform operations of addition and scalar multiplication to obtain vectors of the form a + b, a – b, 5a + 3b, 2a – 7b, etc. Such vectors are called linear combinations of a and b. In general, for any scalars λ and μ, the vector $\lambda a + \mu b$ is a linear combination of a and b.

Thus, if $\{a_1, a_2, \dots, a_k\}$ is a set of k vectors from R^n and $\{\lambda_1, \dots, \lambda_2, \lambda_k\}$ is a set of k scalars, then the vector

$$\lambda_1 a_1 + \lambda_2 a_2 + \dots + \lambda_k a_k$$

is known as a linear combination of the given set of vectors.

$$a_1, a_2, \dots, a_k.$$

Definition 1: *(Linear dependence and independence of vectors). A set of vectors $a_1, a_2, \dots, a_k$ of R^n, is said to be Linearly dependent if there exist scalars $\lambda_1, \lambda_2, \dots, \lambda_k$ not all zero, such that*

$$\lambda_1 a_1 + \lambda_2 a_2 + \dots + \lambda_k a_k = 0.$$

When it is not possible to have the above equality without all l_i's being zero, then the given set of vectors $a_1, a_2, \dots, a_k$ is said to be Linearly Independent.

Illustration 1: The vectors $a_1 = [1, 2]$ and $a_2 = [2, 4]$ are linearly dependence since there exist $\lambda_1 = 2$ and $\lambda_2 = -1$ for which

$$\lambda_1 a_1 + \lambda_2 a_2 = 0.$$

Illustration 2: The vectors $c_1 = [1, 0, 0]$, $c_2 = [0, 1, 0]$ and $c_3 = [0, 0, 1]$ are linearly independent since

$$\lambda_1 c_1 + \lambda_2 c_2 + c_3 \lambda_3 = 0$$

$$\Rightarrow \quad \lambda_1 [1, 0, 0] + \lambda_2 [0, 1, 0] + [0, 0, 1] = [0, 0, 0]$$

$$\Rightarrow \quad [\lambda_1, \lambda_2, \lambda_3] = [0, 0, 0]$$

$$\Rightarrow \quad \lambda_1 = 0, \lambda_2 = 0 \text{ and } \lambda_3 = 0,$$

Note: If a set of vectors is linearly independent, then any sudsier of it is also linearly independent. If a set of vectors is linearly dependent, then a superset of it is also linearly dependent.

Definition 2: *(Basis set).* A linearly independent set of vectors $a_1, a_2, \dots a_n$, of R^n, having the property that any vector of R^n can be expressed as a linear combination of these vectors, is called *Basis set* for R^n.

Definition 3: *(Standard basis).* The set of vectors e_i, $i = 1, 2, \dots n$, is called the *standard basis* for R^n.

Example: *If X is the set of eight vertices of a cube, then the convex hull C(X) is the whole cube.*

Definition 4: *(Simplex). A Simplex in n dimension is a convex polyhedron having exactly (n + l) vertices.*

A simplex in zero dimension is a point, in one dimension it is a fine segment, in two dimensions it is a triangle, in three dimensions it is a tetrahedron and so on.

CONVEX POLYHEDRON

The set of all convex combinations of finite number of points is called the convex polyhedron generated by these points.

Alternatively, if the set X consists of a finite number of points, the convex hull of X is called a convex polyhedron with vertices at these points.

Example: *The set of the area of a triangle is a convex polyhedron of the set of its vertices.*

SOME IMPORTANT THEOREMS

Theorem 1:

Every basic feasible solution of the system $Ax = b$, $x \geq 0$ is an extreme point of the convex set of feasible solutions and conversely.

Proof:

To prove that every B.F.S. is an extreme point of the convex set of all feasible solution.

Let x be a B.F.S. of Ax = b which is a n-component vector containing both zero (non-basic) and non-zero (basic) variables. Let x_B and B be the vector of m basic variables and the matrix of vectors associated to basic variables in the B.F.S. x respectively, then

$$x = [x_B, 0], \qquad ...(1)$$

where 0 is a null vector of (n – m) components,

and $$Ax = b \Rightarrow B.\, x_B = b. \qquad ...(2)$$

Now we have to prove that x is an extreme point.

We shall prove this by using contradiction.

Suppose that x is not an extreme point. If X is the convex set of all feasible solutions of Ax = b, then $x \in X$.

If x is not an extreme point then there exist two distinct points x_1 and x_2 in X such that

$$x = \lambda x_1 + (1 - \lambda)\, x_2,\ 0 < \lambda < 1 \qquad ...(3)$$

But x_1 and x_2 can be expressed as

$$x_1 = [u_1, v_1] \text{ and } x_2 = [u_2, v_2] \qquad ...(4)$$

where u_1 and u_2 are vectors of m components of x_1 and x_2 respectively and v_1, v_2 are (n – m) component vectors.

Substituting the values of x and x_1, x_2 from (1) and (4) in (3), we have

$$[x_B, 0\} = \lambda\, [u_1, v_1] + (1 - \lambda)\, [u_2, v_2], = < \lambda < 1$$

$$= [\lambda u_1 + (1 - \lambda)\, u_2, \lambda v_1 + (1 - \lambda)\, v_2]$$

$$\therefore 0 = \lambda v_1 + (1 - \lambda)\, v_2,\ 0 < \lambda < 1 \qquad ...(5)$$

Now $1 > \lambda > 0$, $1 - \lambda > 0$, and the components of v_1 and v_2 are ≥ 0. The relation (5) can only be satisfied when $v_1 = 0$ and $v_2 = 0$

$\therefore \quad x_1 = [u_1, 0], x_2 = [u_2, 0].$

Since x_1 and x_2 are in X, therefore from (2) we have

$$A\,x_1 = B\,u_1 = b$$

and $$A\,x_2 = B\,u_2 = b$$

i.e., $$b = B\,x_B = B\,u_1 = B\,u_2$$

which gives $x_B = u_1 = u_2$

$\therefore \quad x = x_1 = x_2$

which is contradiction to the assumption that $x_1 \neq x_2$

i.e., x cannot be expressed as a convex combination of any two distinct points in the set of all feasible solutions. Hence x must be an extreme point.

Converse. *To prove that every extreme point of the convex set of feasible solutions is a B.F.S.*

Let $x = [x_1, x_2, \ldots, x_n]$ be an extreme point. Now in order to prove that x is a B.F.S. we shall prove that the vectors associated with the positive elements of x are L.I. Suppose that k-components (variables) in x are non-zero and $(n - k)$ components are zero.

We can assume these components as the first k components of x.

$$\therefore \sum_{i=1}^{n} x_i \alpha_i = b, \; x_i > 0, \; i = 1, 2, \ldots, k. \qquad \ldots(6)$$

where α_i is the column vector in A associated to the ith variables in x.

If possible, let the column vectors $a_1, a_2, \ldots, a_k$ of matrix A be L.D. Then there exist some scalars l_i $(i = 1, 2, \ldots, k)$ with at least one of them non-zero, s.t.,

$$\sum_{i=1}^{n} \lambda_i \alpha_i = 0 \qquad \ldots(7)$$

From (6) and (7), for some arbitrary $\delta > 0$, we have

$$\sum_{i=1}^{k} x_i \alpha_i \pm \delta \sum_{i=1}^{k} \lambda_i \alpha_i = b$$

or $$\sum_{i=1}^{n} (x_i \pm \delta \lambda_i) \alpha_i = b$$

from which it is obvious that the two points

(n – k) in number

$$x_1^* = [x_1 + \delta\lambda_1, x_2 + \delta\lambda_2, \ldots, x_k + \delta\lambda_k, 0, 0, \ldots, 0]$$
$$x_2^* = [x_1 - \delta\lambda_1, x_2 - \delta\lambda_2, \ldots, x_k - \delta\lambda_k, 0, 0, \ldots, 0]$$

satisfy $A x = b$.

Also since $x_i > 0$, therefore, taking δ s.t.

$$0 < d < \underset{i}{\text{Min.}} \left\{ \frac{x_i}{|\lambda_i|} \right\}, \lambda_i \neq 0, i = 1, 2, \ldots, k.$$

We conclude that first k components of x_1^* and x_2^* are always positive. But the remaining components of x_1^* and x_2^* are zero, which follows that x_1^* and x_2^* are feasible solutions different from x.

Now $x_1^* + x_2^* = 2\ [x_1, x_2, \ldots, x_k, 0, 0, \ldots, 0]$

or $$\frac{1}{2} x_1^* + \frac{1}{2} x_2^* = [x_1, x_2, \ldots x_k, 0, 0, \ldots, 0] = x$$

or $$x = \lambda x_1^* + (1 - \lambda)\ x_2^* \text{ where } \lambda = \frac{1}{2}$$

i.e., x can be expressed as a convex combination of two distinct feasible solutions x_1^* and x_2^*.

But this is a contradiction as x is an extreme point. Hence, the vectors $a_1, a_2, \ldots a_k$ are L.I. Further, we know that at most m vectors of E_m can be L.I. So $a_1, a_2, \ldots a_k$ cannot be more than m and hence the extreme point x will have almost m non-zero variables, *i.e.*, atleast $(n - m)$ variables will be zero.

Thus, x is a B.F.S. Hence, every extreme points of the convex set of feasible solutions is a B.F.S.

Cor 1: *The extreme points of the convex set of feasible solutions are finite in number.*

From the above theorem, we conclude that there is only one extreme point for a given B.F.S. and vice versa. That is there is one-to-one correspondence between the extreme points and the B.F. solutions in the absence of degeneracy. Also in case of degeneracy corresponding to an extreme point with the number of non-zero variables less than m, we can form more than one degenerate B.F.S. Hence the number of extreme points of the feasible region is finite and it cannot exceed the number of its B.F. solutions.

Cor 2: *An extreme point can have almost m-positive x_i's where m is the number of constraints.*

Cor 3: *In an extreme point, vectors associated to the positive x_i's are L.I.*

Theorem 2:

If the convex set of the feasible solutions of $Ax = b$, $x \geq 0$ is a convex polyhedron, then atleast one of the extreme points gives an optimal solution.

Proof:

In the Cor. 1 of last theorem we have proved that the extreme points of the convex set of feasible solutions of $Ax = b$, $x \geq 0$ are finite in number.

Let $x_1, x_2, \ldots, x_k$ be the extreme points of the set X of all the feasible solutions of $Ax = b$, $x \geq 0$. Let Z be the objective function which is to be maximized be given by

$$Z = c\,x.$$

If $x^* \in X$ is the optimal solutions, then

$$\text{Max. } Z = c\,x^*.$$

Now if x^* is an extreme point, them the theorem is proved.

Now if x^* is not an extreme point in X, then since X is convex polyhedron, therefore x^* can be expressed as a convex combination of the extreme points of X,

i.e., $$x^* = \lambda_1 x_1 + \lambda_2 x_2 + \ldots + \lambda_k x_k = \sum_{i=1}^{k} \lambda_i . x_i,$$

$$l_i \geq 0 \text{ and } \Sigma\lambda_i = 1$$

$$\therefore Z^* = cx^* = c\,(\lambda_1 x_1 + \lambda_2 x_2 + \ldots + \lambda_k x_k)$$

$$= (\lambda_1 cx_1 + \lambda_2 cx_2 + \ldots + \lambda_k cx_k)$$

If maximum of sx_i is cx_p, then

$$Z^* \leq (\lambda_1 + \lambda_2 + \ldots + \lambda_k).\ cx_p$$

or $$Z^* \leq cx_p.$$

But Z^* is the maximum value of Z. Therefore,

$$Z^* = cx_p \text{ or } cx^* = cx_p$$

i.e., $$x^* = x_p \text{ (one of the extreme points).}$$

Hence the optimal solution is attained at the extreme point.

This proves the theorem.

Theorem 3:

If the objective function of a L.P.P. assumes its optimal value at more than one extreme point, then every convex combination of these extreme points gives the optimal value of the objective function.

Proof:

Let us consider the L.P.P. as follows:

Max. $Z = cx$

s.t. $Ax = b, x \geq 0.$

Let $x_1, x_2, \ldots x_k$ be the extreme points of the feasible region. If the objective function Z assumes its optimal value Z* at the extreme points $x_1, x_2, \ldots, x_p$, $(p \leq k)$ then

$$Z^* = cx_1 = cx_2 = \ldots = cx_p.$$

If x_0 is the convex combination of the extreme points $x_1, x_2, \ldots x_p$, then

$$x_0 = \lambda_1 x_1 + \lambda_2 x_2 + \ldots + \lambda_p x_p, \lambda_i \geq 0, \sum_{i=1}^{p} \lambda_i = 1$$

$$\therefore cx_0 = c[\lambda_1 x_1 + \lambda_2 x_2 + \ldots + \lambda_p x_p]$$

$$= \lambda_1 cx_1 + \lambda_2 cx_2 + \ldots + \lambda_p cx_p$$

$$= \lambda_1 Z^* + \lambda_2 Z^* + \ldots + \lambda_p Z^*$$

$$= (\lambda_1 + \lambda_2 + \ldots + \lambda_p) Z^* = Z^*, \qquad \because \sum_{i=1}^{p} \lambda_i = 1.$$

Hence the optimal value Z* is also attained at x_0 which is the convex combination of the extreme points at which optimal value occurs. Hence the theorem.

Theorem 4:

A hyperplane is a convex set.

Proof:

Consider the hyperplane

$$X = \{x : c\,x = z\}.$$

Let x_1 and x_2 be any two points in the hyperplane X.

$\therefore c\,x_1 = z$ and $c\,x_2 = z.$

If $x_3 = \lambda x_1 + (1 - \lambda) x_2, 0 \leq \lambda \leq 1$

then $cx_3 = \lambda c\,x_1 + (1 - \lambda) cx_2,$

$$= \lambda z + (1 - \lambda) z = z$$

which implies that

$x_3 = \lambda x_1 + (1 - \lambda) x_2$ is also a point in the hyperplane X.

Hence by definition, the hyperplane X is a convex set.

Theorem 5:

The closed half spaces $H_1 = \{x : c\,x \geq z\}$ and $H_2 = \{x : c\,x \leq z\}$ are convex sets.

Proof:

Let x_1 and x_2 be any two points of H_1. Then $cx_1 \geq z$, $cx_2 \geq z$.

If $0 \leq 1 \leq 1$, then $c\,[\lambda x_1 + (1 - \lambda) x_2] = \lambda\, cx_1 + (1 - \lambda)\, cx_2$

$\geq \lambda z + (1 - \lambda) z = z.$

Hence $x_1, x_2 \in H_1$ and $0 \leq 1 \leq 1$

$\Rightarrow \quad \lambda x_1 + (1 - \lambda) x_2 \in H_1.$

So H_1 is a convex set.

Similarly, if $x_1, x_2 \in H_2$, $0 \leq \lambda \leq 1$, then replacing the inequality sign $\geq$ by $\leq$ in above, we get st$x_1 + (1 - \lambda) x_2 \in H_2$.

So H_2 is also a convex set.

Theorem 6:

The set of n vector e_i, $i = 1, 2, \ldots, n$ where e_i has i th component equal to 1, and all other components zeroes, forms a basis for R^n.

Proof:

We shall show that the vectors $e_1, e_2, \ldots, e_n$ of R^n are linearly independent and that any vector of R^n can be expressed as linear combination of these unit vectors.

Consider $\sum_{i=1}^{n} \lambda_i e_i = 0$, where λ_i $(i = 1, 2, \ldots n)$ are scalars,

or $\quad 11\,[1, 0, 0, \ldots 0] + \lambda_2\,[0, 1, 0, \ldots, 0] + \ldots + \lambda_n\,[0, 0, \ldots, 1] = 0$

or $\quad [\lambda_1, \lambda_2, \lambda_3, \ldots, \lambda_n] = [0, 0, 0, \ldots, 0].$

This implies that $\lambda_i = 0$ for all $i = 1, 2, \ldots n$.

Thus, the n vectors e_i, $i = 1, 2, \ldots, n$, are linearly independent.

Let x be any other vector of R^n different from the vectors

e_i, $i = 1, 2, \ldots, n$.

This can be written as

$$x = [x_1, x_2, .. x_k, ... ,x_n]$$
$$= x_1\ e_1 + x_2\ e_2 + ... + x_n\ e_i$$
$$= \sum_{i=1}^{n} x_i e_i$$

This happens to be a linear combination of the n unit vectors e_i; i = 1, 2, ... ,n. Since x is any vector of R^n different from the n unit vectors, therefore, every vector of R^n can be expressed as a linear combination of these vectors.

Hence {e, i = 1, 2, ... ,n} forms a basis for R^n.

Theorem 7:

(Replacement Theorem). *Let $\{a_1, a_2, ... ,a_n\}$ be any basis set for R^n and $b \neq 0$ any other vector in R^n. If b is expressible as*

$$b = \sum_{i=1}^{n} \delta_i a_i \quad \text{where } d_i \neq 0, \text{ for some } i,$$

then the vector ak (for which $\delta_k \neq 0$) can be replaced by the vector b. The new set of n vectors $a_1, a_2, ... , a_{k-1}, a_{k-1}, ... ,a_n$, b is also a basis set for R^n.

Proof:

Let $\delta_k \neq 0$. Consider the new set of vectors

$$a_1, a_2, ... , a_{k-1}, a_{k+1}, ... ,a_n, b.$$

To prove that this set is a basis for R^n, we must show

(i) $a_1, a_2, ... , a_{k-1}, a_{k+1}, a_n$, b are linearly independent, and

(ii) any vector of R^n can be expressed as a linear combination of the vectors $a_1, a_2, ... , a_{k-1}, a_{k+1}, ... ,a^n$, b.

Let us suppose the set $\{a_1, a_2, ... ,a_n, ... b\}$ 10 net a linearly independent set.

Then the equation $\sum_{i=j}^{n} \lambda_i a_i + \lambda_b = 0$

will hold, even with some non-zero scalars

$$\lambda i\ [i = 1, 2, ... ,k - 1, k + 1 ... n]$$

now $\lambda = 0$ implies

$$\sum_{i=1}^{n} \lambda_i a_i = 0 \text{ with bow } \lambda_i \neq 0.$$

This contradicts the feat third the set $[a_1, a_2, ... a_{ks}, a_{k+1} ... a_n]$ is linearly independent set,

Thus $l \neq 0$ amul we have

$$\sum_{i=1}^{n} \lambda_i a_i + \lambda \sum_{i=1}^{n} \delta_i a_i = 0 \left[\text{since } b = \sum_{i=1}^{n} \delta_i a_i\right]$$

$$\sum_{i=1 i \neq k}^{n} (\lambda i + \lambda \delta i) ai + \lambda a\, \delta ak = 0$$

Since $l \neq 01$ $d_u \neq c$ the vector the set of vectors $a_1, a_2 \ldots a_n$ is linearly dependent. This centralised the hypothesis.

Thus $\{a_1, a_2 \ldots a_{k-1}, a_{k+1}, \ldots a_n, b]$ is a linearly in dependent set

Now let x be any vector in R^n other thou the given basis vectors $a_1, a_2 \ldots a_n$ then we can express x as

$$n = \sum_{i=1}^{n} \lambda_i a_i$$ where λ_i [i = 1, 2, ... ,n] are scalars.

$$\text{Now since } b = \sum_{i=1}^{n} \delta_i a_i = \sum_{i=1 i=k}^{n} \delta_i a_i + \delta_k a \quad , \delta_1 \neq 0$$

$$\text{we have } a_u = \frac{1}{\delta u}\left[b - \sum_{i=1, i \neq k}^{n} \delta_i a_i\right]$$

$$= \frac{b}{\delta k} - \sum_{i=1, i \neq k}^{n} \frac{bi}{\delta k} a_i$$

$$x = \sum_{i=1, i \neq k}^{n} \lambda_i a_i + \lambda_k \left[\frac{b}{\delta_k} - \sum_{i=1, i \neq k}^{n} \frac{\delta_i}{\delta_k} a_i\right]$$

$$= \sum_{i=1, i \neq k}^{n} \left(\lambda_1 - \frac{\lambda_k}{\delta_k} \delta_i\right) a_i + \frac{\lambda_k}{\delta_k} b.$$

Thus, x can be expressed as a linear combination of the vectors $a_1, a_2, \ldots, a_{k-1}, a_{k+1}, a_n$, b.

Hence the set $\{a_1, a_2, \ldots, a_{k-1}, a_{k+1}, \ldots, a_n, b\}$ forms a basis for R_n.

Theorem 8:

Intersection of two convex sets is also a convex set.

Proof:

Consider tow convex sets X_1 and X_2. Let X_3 be the intersection of sets X_1 and X_2, *i.e.*, $X_3 = X_1 \cap X_2$.

Now $\quad x_1 \in X_1 \cap X_1$

$\Rightarrow \quad x_1 \in X_1$ and $x_1 \in X_2$

$x_2 \in X_1 \cap X_2$

$\Rightarrow \qquad x_2 \in X_1$ and $x_2 \in X_2$.

Since X_1 and X_2 are convex sets,

$\therefore \qquad x_1, x_2 \in X_1 \Rightarrow \lambda x_1 + (1 - \lambda)\, x_2 \in X_1 \qquad 0 \le l \le 1$

and $\qquad x_1, x_2 \in X_2 \Rightarrow \lambda x_1 + (1 - \lambda)\, x_2 \in X_2 \qquad 0 \le l \le 1$

Thus $\lambda x_1 + (1 - \lambda)\, x_2 \in X_1$ and $\lambda x_1 + (1 - \lambda)\, x_2 \in X_2$

$\Rightarrow \lambda x_1 + (1 - \lambda)\, x_2 \in X_1 \cap X_2 \qquad 0 \le l \le 1.$

Hence by definition, $X_3 = X_1 \cap X_2$ is a convex set.

Theorem 9:

The set of all convex combinations of a finite number of points $x_1, x_2, \ldots, x_n$ is a convex set.

Proof:

Let X be the set of all convex combinations of a finite number of point.

i.e., $\qquad X = \left\{ x : x = \sum_{i=1}^{n} \lambda_i x_i, \sum_{i=1}^{n} \lambda_i = 1, \lambda_i \ge 0 \right\}.$

Let $\qquad u, v \in X$

$$\therefore u = \sum_{i=1}^{n} a_i x_i ; \sum_{i=1}^{n} a_i = 1,\ a_i \ge 0$$

and $$v = \sum_{i=1}^{n} b_i x_i \sum_{i=1}^{n} b_i = 1,\ b_i \ge 0.$$

Consider $w = \lambda u + (1 - \lambda)\, v,\ 0 \le \lambda \le 1$

$$\therefore w = !\ \sum_{i=1}^{n} a_i x_i + (1-\lambda) \sum_{i=1}^{n} b_i x_u$$

$$= \sum_{i=1}^{n} \{\lambda a_i + (1-\lambda) b_i\} x_i$$

$$= \sum_{i=1}^{n} c_i x_i \text{ where } c_i = \lambda a_i + (1 - \lambda)\, b_i$$

Now $$\sum_{i=1}^{n} c_i = \sum_{i=1}^{n} \{\lambda a_i + (1-\lambda) b_i\}$$

$$= 1 \sum_{i=1}^{n} a_i + (1-\lambda) \sum_{i=1}^{n} b_i = \lambda.1 + (1 - \lambda).\ 1 = 1.$$

Also $c_i = \lambda a_i + (1 - \lambda)\, b_i \ge 0,\ V$ i.

Hence $w = \sum_{i=1}^{n} c_i x_i$ is a convex combination of $x_1, x_2, \ldots, x_n$.

i.e., $w = \lambda u + (1 - \lambda) v \in X, 0 \leq l \leq 1.$

Hence by definition X is a convex set.

Theorem 10:

Let S and T be two convex sets in E^n, then for any scalars α, β, $\alpha S + \beta T$ is also convex.

Proof:

Let x, y be two points of $\alpha S + \beta T$.

Then $x = \alpha u_1 + \beta v_1$ and $y = \alpha u_2 + \alpha v_2$ ...(1)

where $u_1, u_2 \in S$ and $v_1, v_2 \in T$.

For any scalar l, $0 \leq l \leq 1$, we have

$$\lambda x + (1 - \lambda) y = \lambda (\alpha u_1 + \beta v_1) + (1 - \lambda) (\alpha u_2 + \beta v_2)$$

$$= \alpha [\lambda u_1 + (1 - \lambda) u_2] + \beta [\lambda v_1 + (1 - \lambda) v_2] \quad ...(2)$$

But S and T are convex sets,

$u_1, u_2 \in S$

$\Rightarrow \quad \lambda u_1 + (1 - \lambda) u_2 \in S, 0 \leq \lambda \leq 1$ ...(3)

and $\quad v_1, v_2 \in T$

$\Rightarrow \quad \lambda v_1 + (1 - \lambda) v_2 \in T, 0 \leq \lambda \leq 1.$ ...(4)

Now from (2), (3) and (4) we have

$$\lambda x + (1 - {}_l) y \in \alpha S + \beta T, 0 \leq l \leq 1.$$

Thus $x, y \in \alpha S + \alpha T$

$\Rightarrow \quad [x : y] \subset \alpha S + \beta T.$

Hence $\alpha S + \beta T$ is a convex set.

Corollary:

If S and T be two convex sets in E^n, then S + T and S – T are also convex sets.

Theorem 11:

A set C is convex iff every convex linear combination of points in C also belongs to C.

Proof:

Let every convex linear combination of points in C belong to C.

Then, in particular convex linear combination of every two points in C also belong to C.

Hence C is a convex set.

Conversely let C be a convex set. Then to prove that convex linear combination of any number of points in C also belong to C.

We shall use the induction principle.

Since C is convex, the convex linear combination of two points in C belongs to C. Thus the result is true for n = 2.

Now suppose the convex linear combination of any n points in C, belongs to C.

Let $x_1, x_2, \ldots, x_n$ be any n points in C. Then by assumption, $\lambda_1 x_1 + \lambda_2 x_2 + \ldots + \lambda_n x_n \in C$ where $\lambda_i \geq 0$, $\Sigma\lambda_i = 1$.

Let $x = \mu_1 x_1 + \mu_2 x_2 + \ldots + \mu_n x_n + \mu_{n+1} x_{n+1}$ ...(1)

i.e., x is a convex linear combination of (n + 1) points of C, where $\mu_i \geq 0$ and $\sum_{i=1}^{n+1} \mu_i = 1$.

Now we shall show that $x \in C$.

If $m_{n+1} = 0$ then x becomes a convex linear combination of $x_1, x_2, \ldots, x_n$ which by assumption, belongs to C and hence the result holds in this case. Also if $m_{n+1} = 1$, then the result is trivially true.

Let m_{n+1} be neither 0 nor 1. Then

$$x = \frac{(\mu_1 + \mu_2 + \ldots + \mu_n)}{(\mu_1 + \mu_2 + \ldots + \mu_n)}(\mu_1 x_1 + \mu_2 x_2 + \ldots + \mu_n x_n) + \mu_{n+1} x_{n+1} \quad \ldots(2)$$

$$= \left(\sum_{i=1}^{n} \mu_i\right)\left(\sum_{i=1}^{n} a_i x_i\right) + \mu_{n+1} x_{n+1},$$

where $a_i = \dfrac{\mu_i}{\mu_i + \mu_2 + \ldots + \mu_n}$, i = 1, 2, ... ,n.

As each $\mu_i \geq 0$, we have $a_i \geq 0$

and also $\sum_{i=1}^{n} a_i = \sum_{i=1}^{n}\left(\dfrac{\mu_i}{\mu_1 + \mu_2 + \ldots + \mu_n}\right) = \dfrac{\Sigma\mu_i}{\Sigma\mu_i} = 1.$

Thus $\sum_{i=1}^{n} a_i x_i = y$ (say) is also a convex linear combination of $x_1, x_2, \ldots, x_n$. So y belongs to C, by assumption.

Now $x = (\mu_1 + \mu_2 + \ldots + \mu_n)\, y + \mu_{n+1}\, x_{n+1}$ ...(3)

But $(\mu_1 + \mu_2 + \ldots + \mu_n) \geq 0,\ \mu_{n+1} \geq 0$

and $(\mu_1 + \mu_2 + \ldots + \mu_n) + \mu_{n+1} = 1.$

It follows that x is a convex combination of two points y and x_{n+1} of C. So x belongs to C.

Hence convex linear combination of (n + 1) points of C also belongs to C. So by induction principle the result is true.

Theorem. 12:

The set of all feasible solution (if not empty) of a L.P.P. is a convex set.

Proof:

Let X be the set of all feasible solution of a L.P.P.

Case I. If the set X has only element, then X is convex set. Hence the theorem is true in this case.

Case II. If the set X has at least two elements.

Let x_1 and x_2 be any two distinct elements in X.

$\therefore$ $A\,x_1 = b,\ x_1 \geq 0.$

and $A\,x_2 = b,\ x_2 \geq 0.$

If $x_3 = \lambda x_1 + (1 - \lambda)\, x_2,\ 0 \leq \lambda \leq 1$

then $Ax_3 = A\lambda x_1 + (1 - \lambda)\, Ax_2$

$= \lambda b + (1 - \lambda)\, b = b.$

Also since $x_1 \geq 0,\ 2 \geq 0,\ \lambda \geq \lambda \geq 0$, as $0 \leq \lambda \leq 1$

$\therefore$ $x_3 = \lambda x_1 + (1 - \lambda)\, x_2 \geq 0$

i.e., x_3 satisfies (1). Thus $x_3 = \lambda x_1 + (1 - \lambda)\, x_2$ is also a F.S. and so belongs to set X.

But x_3 is a convex combination of any two distinct points, x_1 and x_2 in X.

Hence by definition the set X is a convex set.

Note: Since the convex combinations of two points are infinite in number so from the above theorem we conclude that:

If a given L.P.P. has two feasible solutions, then it has infinite number of feasible solutions.

Theorem 13:

(a) Intersection of any finite number of convex sets is also a convex set.

Proof:

Let $X_1, X_2, \ldots, X_n$ be n convex sets and

$$X = X_1 \cap X_2 \cap \ldots \cap X_n.$$

Now $\quad x_1 \in X_1 \cap X_2 \cap \ldots \cap X_n$

$\Rightarrow \quad x_1 \in X_i, \forall\ i = 1, 2, \ldots n$

and $\quad x_2 \in X_1 \cap X_2 \cap \ldots \cap X_n$

$\Rightarrow \quad x_2 \in X_i, \forall\ i = 1, 2, \ldots n.$

Since X– is convex set for i = 1, 2, ... ,n

$\therefore \quad x_1, x_2 \in X_i$

$\Rightarrow \quad \lambda x_1 + (1 - \lambda) x_2 \in X_i, \forall\ i = 1, 2, \ldots, n$

where $0 \leq 1 \leq 1$

$\Rightarrow \quad \lambda x_1 + (1 - \lambda) x_2 \in X_1 \cap X_2 \cap \ldots \cap X_n\ 0 \leq \lambda \leq 1.$

i.e., $\quad x_1 \in X_1 \cap X_2 \cap \ldots \cap X_n$

and $\quad x_2 \in X_1 \cap X_2 \cap \ldots \cap X_n$

$\Rightarrow \quad \lambda x_1 + (1 - \lambda) x_2 \in X_1 \cap X_2 \cap \ldots \cap X_n, 0 \leq \lambda \leq 1.$

Hence by definition $X_1 \cap X_2 \cap \ldots \cap X_n$ is a convex set.

(b) *Arbitrary intersection of convex sets is also a convex set.*

SOLVED EXAMPLES

Example 1(a):

Find the extreme points, if any, the following sets:

(i) $S = \{(x, y) : x_2 + y_2 \leq 25\}$

(ii) $\{(x, y) : |x| \leq 1, |y| \leq 1\}$.

Solution:

(i) Draw the region of the set S. It represents the boundary and interior of the circle with centre at (0, 0) and radius 5.

Every point on its circumference is an extreme point.

(ii) We have $|x| \leq 1 \Rightarrow -1 \leq x \leq 1$ and $|y| \leq 1$

$\Rightarrow -1 \leq y \leq 1.$

Thus the set S represents the region of rectangle bounded by the lines $x = 1, x = -1, y = 1, y = -1$.

The extreme points of this convex set are (1, 1), (– 1, 1), (– 1, – 1) and (1, – 1).

Example 1(b):

Is the union of two convex sets, necessarily a convex set?

Solution:

No, it is not necessary that union of two convex sets will be a convex set.

For example: $S_1 = \{(x, y) : x \geq 1\}$, $S_2 = \{(x, y) : x \geq 2\}$

and $T_1 = \{(x, y) : x \geq 1\}$, $T_2 = \{(x, y) : y \geq 1\}$.

Then $S_1 \cup S_2$ is again a convex set while $T_1 \cup T_2$ is not a convex set.

Example 1(c):

Find all the basic feasible solutions for the equations

$$2x_1 + 6x_2 + 2x_3 + x_4 = 3$$

$$6x_1 + 4x_2 + 4x_3 + 6x_4 = 2$$

$$x_i \geq 0$$

and determine the associated general convex combination of the extreme point solutions.

Solution:

In matrix form the given system of equations can be written as

$$A x = b$$

where $A = (\alpha_1, \alpha_2, \alpha_3, \alpha_4)$

$$\alpha_1 = \begin{bmatrix} 2 \\ 6 \end{bmatrix}, \alpha_2 = \begin{bmatrix} 6 \\ 4 \end{bmatrix}, \alpha_3 = \begin{bmatrix} 2 \\ 4 \end{bmatrix}, \alpha_4 = \begin{bmatrix} 1 \\ 6 \end{bmatrix}, x = \begin{bmatrix} x_1 \\ x_2 \\ x_3 \\ x_4 \end{bmatrix}, b = \begin{bmatrix} 3 \\ 2 \end{bmatrix}.$$

This problem can have at most ${}^4C_2 = 6$ basic solutions. Now the six sets of two vectors are

$$B_1 = [\alpha_1, \alpha_2] = \begin{bmatrix} 2 & 6 \\ 6 & 4 \end{bmatrix}, B_2 = [\alpha_1, \alpha_3] = \begin{bmatrix} 2 & 2 \\ 6 & 4 \end{bmatrix}$$

$$B_3 = [\alpha_1, \alpha_4] = \begin{bmatrix} 2 & 1 \\ 6 & 6 \end{bmatrix}, B_4 = [\alpha_2, \alpha_3] = \begin{bmatrix} 6 & 2 \\ 4 & 4 \end{bmatrix}$$

$$B_5 = [\alpha_2, \alpha_4] = \begin{bmatrix} 6 & 1 \\ 4 & 6 \end{bmatrix}, \; B_6 = [\alpha_3, \alpha_4] = \begin{bmatrix} 2 & 1 \\ 4 & 6 \end{bmatrix}.$$

Here $|B_1| = -28, \; |B_2| = -4,$

$|B_3| = 6, \; |B_4| = 16,$

$|B_5| = 32 \; |B_6| = 8.$

Since none of these is zero, therefore all these sets are L.I.

Hence all the six basic solutions exist.

If x_{Bi}, i = 1, 2, ... , 6 are the vectors of the basic variables associated to the sets B_i, i = 1, 2, ... , 6 respectively, then

$$\begin{bmatrix} x_1 \\ x_2 \end{bmatrix} = x_{B1} = B_1^{-1} b = -\frac{1}{28}\begin{bmatrix} 4 & -6 \\ -6 & 2 \end{bmatrix}\begin{bmatrix} 3 \\ 2 \end{bmatrix} = \begin{bmatrix} 0 \\ 1/2 \end{bmatrix}$$

$$\begin{bmatrix} x_1 \\ x_3 \end{bmatrix} = x_{B2} = B_2^{-1} b = -\frac{1}{4}\begin{bmatrix} 4 & -2 \\ -6 & 2 \end{bmatrix}\begin{bmatrix} 3 \\ 2 \end{bmatrix} = \begin{bmatrix} -2 \\ 7/2 \end{bmatrix}$$

$$\begin{bmatrix} x_1 \\ x_4 \end{bmatrix} = x_{B3} = B_3^{-1} b = \frac{1}{6}\begin{bmatrix} 6 & -1 \\ -6 & 2 \end{bmatrix}\begin{bmatrix} 3 \\ 2 \end{bmatrix} = \begin{bmatrix} 8/3 \\ -7/3 \end{bmatrix}$$

$$\begin{bmatrix} x_2 \\ x_3 \end{bmatrix} = x_{B4} = B_4^{-1} b = \frac{1}{16}\begin{bmatrix} 4 & -2 \\ -4 & 6 \end{bmatrix}\begin{bmatrix} 3 \\ 2 \end{bmatrix} = \begin{bmatrix} 1/2 \\ 0 \end{bmatrix}$$

$$\begin{bmatrix} x_2 \\ x_4 \end{bmatrix} = x_{B5} = B_5^{-1} b = \frac{1}{32}\begin{bmatrix} 6 & -1 \\ -4 & 6 \end{bmatrix}\begin{bmatrix} 3 \\ 2 \end{bmatrix} = \begin{bmatrix} 1/2 \\ 0 \end{bmatrix}$$

$$\begin{bmatrix} x_3 \\ x_4 \end{bmatrix} = x_{B6} = B_6^{-1} b = \frac{1}{8}\begin{bmatrix} 6 & -1 \\ -4 & 2 \end{bmatrix}\begin{bmatrix} 3 \\ 2 \end{bmatrix} = \begin{bmatrix} 2 \\ -1 \end{bmatrix}.$$

Thus it is obvious that out of these only three basic solutions are B.F.S. (in which variables) are non-negative). But the B.F.S.'s correspond to the extreme points. Hence the only three extreme point solutions are given by

$$x_1 = \left(0, \frac{1}{2,}, 0, 0\right),$$

$$x_2 = \left(0, \frac{1}{2}, 0, 0\right),$$

$$x_3 = \left(0, \frac{1}{2}, 0, 0\right).$$

Here $x_1 = x_2 = x_3$. Hence there is unique extreme point solution.

Note: To find the basic solution x_{B1} we can also proceed as follows.

Putting $x_3 = 0$, $x_4 = 0$ in the given eqns., we get $2x_1 + 6x_2 = 3$ and $6x_1 + 4x_2 = 2$ solving $x_1 = 0$, $x_2 = 1/2$, etc.

Example 2(a):

Show that $C = \{(x_1, x_2) : 2x_1 + 3x_2 = 7\} \subset R^2$ is a convex set.

Solution:

Let u, v $\in$ C, where $u = (u_1, u_2)$, $v = (v_1, v_2)$.

Then $2u_1 + 3u_2 = 7$ and $2v_1 + 3v_2 = 7$.

If $w = (w_1, w_2)$ is a point on the line segment joining the points u and v, then

$$w = \lambda u + (1 - \lambda) v, 0 \le 1 \le 1$$

$$\Rightarrow (w_1, w_2) = \lambda (u_1, u_2) + (1 - \lambda) (v_1, v_2)$$

$$= (\lambda u_1 + (1 - \lambda) v_1, \lambda u_2 + (1 - \lambda) v_2)$$

$$w_1 = \lambda u_1 + (1 - \lambda) v_1 \text{ and } w_2 = \lambda u_2 + (1 - \lambda) v_2.$$

$$2w_1 + 3w_2 = 2 [\lambda u_1 + (1 - \lambda) v_1] + 3 [\lambda u_2 + (1 - \lambda) v_2]$$

$$= \lambda [2u_1 + 3u_2] + (1 - \lambda) [2v_1 + 3v_2]$$

$$= \lambda. 7 + (1 - \lambda). 7 = 7, \text{ using (1)}$$

$$\therefore w = (w_1, w_2) \in C.$$

Hence the set C is a convex set.

Example 2(b):

Show that $S = \{(x_1, x_2, x_3) : 2x_1 - x_2 + x_3 \le 4\} \subset R^3$, is a convex set.

Solution:

Let $x = (x_1, x_2, x_3)$ and $y = (y_1, y_2, y_3)$ be any two point of S. Then we have

$$2x_1 + - x_2 + x_3 \le 4 \text{ and } 2y_1 - y_2 + y_3 \le 4 \quad ...(1)$$

If $w = (w_1, w_2, w_3)$ is a point on the line segment joining the points x and y, then

$$w = \lambda x + (1 - \lambda) y, 0 \le 1 \le 1$$

$$\Rightarrow (w_1, w_2, w_3) = \lambda (x_1, x_2, x_3) + (1 - \lambda) (y_1, y_2, y_3)$$

$$\Rightarrow w_1 = \lambda x_1 + (1 - \lambda) y_1, w_2 = \lambda x_2 + (1 - \lambda) y_2, w_3 = \lambda x_3 + (1 - \lambda) y_3.$$

Now $2w_1 - w_2 + w_3 = \lambda (2x_1 - x_2 = x_3) + (1 - \lambda) (2y_1 - y_2 + y_3)$

$$\le 4\lambda + 4 (1 - \lambda), \text{ using (1)}$$

$$= 4$$

$\therefore\ w = (w_1, w_2, w_3) \in S.$

Hence the set S is a convex set.

Example 3(a):

Examine convexity to the set

$\{(x_1, x_2) \in R^2 : 4x_1 + 3x_2 \leq 6,\ x_1 + x_2 \geq 1\}$

Solution:

Let $S = \{(x_1, x_2) \in R^2 : 4x_1 + 3x_2 \leq 6$ and $x_1 + x_2 \geq 1$

If $u = (x_1, x_2) \in S$ Then $4x_1 + 3x_2 \leq 6$ and $x_1 + x_2 \geq 1$

and if $v = (y_1, y_2) \in S$ then $4y_1 + 3y_2 \geq 6$ and $x_1 + y_2 \geq 1$.

If $w = (z_1, z_2)$ is a point on the line segment joining points u and v, then

$$w = \lambda u + (1 - \lambda)\ v \qquad 0 \leq 1 \leq 1$$

$\Rightarrow (z_1, z_2) = \lambda\ (x_1, x_2) + (1 - \lambda)\ (y_1, y_2)$

$= (\lambda x_1 + (1 - \lambda)\ y_1, \lambda x_2 + (1 - \lambda)\ y_2)$

$\Rightarrow z_1 = \lambda x_1 + (1 - \lambda)\ y_1$ and $z_2 = \lambda x_2 + (1 - \lambda)\ y_2$

Now $4z_1 + 3z_2 = \lambda\ (4x_1 + 3x_2) + (1 - \lambda)\ (4y_1 + 3y_2) \leq 6^*$

$\because 0 \leq \lambda \leq 1,\ 0 \leq 1 - \lambda \leq 1,\ 4x_1 + 3x_2 \leq 6$ and $4y_1 + 3y_2 \leq 6$.

and $\qquad z_1 + z_2 = \lambda\ (x_1 + x_2) + (1 - \lambda)\ (y_1 + y_2) \geq 1^{**}$

$\because 0 \leq \lambda \leq 1,\ 0 \leq 1 - \lambda \leq 1,\ x_1 + x_2 \geq 1,\ y_1 + y_2 > 1.$

$\because w = (z_1, z_2) \in S.$

Hence the set S is a convex set.

Note: As the extreme case if $4x_1 + 3x_2 = 6$, $4y_1 + 3y_2 = 6$,

then $4z_1 + 3z_2 = \lambda.\ 6 + (1 - \lambda)\ 6 = 6$

and if $x_1 + x_2 = 1$, $y_1 + y_2 = 1$

then $z_1 + z_2 = \lambda.\ 1 + (1 - \lambda).\ 1 = 1.$

Aliter: Obviously S is the intersection of two half spaces, viz.,

$$H_1 = \{(x_1, x_2) : 4x_1 + 3x_2 \leq 6\}$$

and $$H_2 = \{(x_1, x_2) : x_1 + x_2 \geq 1\}.$$

Since H_1 and H_2 are convex sets so $S = H_1 \cap H_2$ is also convex.

Example 3(b):

Show that the set $S = \{x : x = (x_1, x_2, x_3),\ x_1^2 + x_2^2 + x_3^2 \leq 1\}$ *is a convex set.*

Solution:

Let x, y ∈ S, where $x = (x_1, x_2, x_3)$, $y = (y_1, y_2, y_3)$.

Then, by the given condition, we have

$$x_1^2 + x_2^2 + x_3^2 \le 1 \quad ...(1)$$

and $$y_1^2 + y_2^2 + y_3^2 \le 1. \quad ...(2)$$

If $z = (z_1, z_2, z_3)$ is a point on the line segment joining the points x and y, then

$$z = \lambda x + (1 - \lambda) y, \; 0 \le \lambda \le 1$$

$$\Rightarrow (z_1, z_2, z_3) = 1 (x_1, x_2, x_3) + (1 - \lambda) (y_1, y_2, y_3)$$

$$\Rightarrow z_1 = \lambda x_1 + (1 - \lambda) y_1, \; z_2 = \lambda x_2 + (1 - \lambda) y_2, \; z_3 = \lambda x_3 + (1 - \lambda) y_3.$$

Now $z_1^2 + z_2^2 + z_3^2 = [\lambda x_1 + (1 - \lambda) y_1]^2 + [\lambda x_2 + (1 - \lambda) y_2]^2 + \lambda x_3 + (1 - \lambda) y_3]^2$

$= \lambda_2 (x_1^2 + x_2^2 + x_3^2) + (1 - \lambda)^2 (y_1^2 + y_2^2 + y_3^2) + 2\lambda (1 - \lambda) (x_1 y_1 + x_2 y_2 + x_3 y_3)$

$\le \lambda_2 . 1 + (1 - \lambda)^2 . \lambda + 2_1 (1 - \lambda) (x_1 y_1 + x_2 y_2 + x_3 y_3)$, using (1) and (2)

By Lagrange's identity, we have

$(x_1^2 + x_2^2 + x_3^2) (y_1^2 + y_2^2 + y_3^2) - x_1 y_1 + x_2 y_2 + x_3 y_3)^2 = \Sigma (x_1 y_2 - x_2 y_1)^2 \ge 0$

$\therefore x_1 y_1 + x_2 y_2 + x_3 y_3 \le \sqrt{}(x_1^2 + x_2^2 + x_3^2) \sqrt{}(y_1^2 + y_2^2 + y_3^2) \le 1$, using (1) and (2)

Thus $z_1^2 + z_2^2 + z_3^2 \le \lambda_2 + \lambda_2 + (1 - \lambda)^2 + 2_1 . (1 - \lambda) = 1.$

$\therefore z = (z_1, z_2, z_3) \in S.$

Hence the set S is a convex set.

Example 3(c):

For any points $x, y \in R^n$, show that the line segment [x : y] is a convex set.

Solution:

Let $u, v \in [x : y]$. Then

$$u = \lambda_1 x + (1 - \lambda_1) y, \; 0 \le \lambda_1 \le 1$$

and $$v = \lambda_2 x + (1 - \lambda_2) y, \; 0 \le \lambda_2 \le 1.$$

Now let w be a point on the line segment joining the points u and v.

Then $w = \lambda u + (1 - \lambda) v, \; 0 \le \lambda \le 1.$

$$= \lambda [\lambda_1 x + (1 - \lambda_1) y] + (1 - \lambda) [\lambda_2 x + (1 - \lambda_2) y]$$
$$= [\lambda\lambda_1 + (1 - \lambda) \lambda_2] x + [\lambda (1 - \lambda_1) + (1 - \lambda) (1 - \lambda_2)] y.$$

If we take $\mu = \lambda\lambda_1 + (1 - \lambda) \lambda_2$, then

$$1 - \mu = 1 - \lambda\lambda_1 - (1 - \lambda) \lambda_2 = \lambda + (1 - \lambda) - \lambda\lambda_1 - (1 - \lambda) \lambda_2$$
$$= \lambda (1 - \lambda_1) + (1 - \lambda) (1 - \lambda_2).$$

Since, $0 \leq \lambda_1 \leq 1, 0 \leq \lambda_2 \leq 1$

$\Rightarrow 0 \leq \lambda\lambda_1 + (1 - \lambda) \lambda_2 \leq 1,$

$$w = \mu x + (1 - \mu) y, 0 \leq \mu \leq 1$$

$\therefore w \in [x : y].$

Hence the set $[x : y]$ is a convex set.

Example 4(a):

A hyperplane is given by the equation

$$3x_1 + 2x_2 + 4x_3 + 7x_4 = 8.$$

and in which half spaces do the following points (– 6, 1, 7, 2) and (1, 2, – 4, 1) lie.

Solution:

The given equation of the hyperplane is

$$3x_1 \mid 2x_2 + 4x_3 + 7x_4 = 8.$$

Substituting (– 6, 1, 7, 2 in the L.H.S., we get

L.H.S. $= 3 (- 6) + 2. 1 + 4. 7 + 7. 2 = 26 > 8 =$ R.H.S.

$\therefore$ the point (– 6, 1, 7, 2) lies in the open half space,

$$3x_1 + 2x_2 + 4x_3 + 7x_4 > 8.$$

Ageing substituting (1, 2, – 4, 1) in the L.H.P., we get

L.H.S. $= 3. 1 + 2. 2 + 4 (- 4) + 7. 1 = - 2 < 8 =$ R.H.S.

$\therefore$ the point (1, 2, – 4, 1) lies in the open half space

$$3x_1 + 2x_2 + 4x_3 + 7x_4 < 8.$$

Example 4(b):

Express any point w inside a triangle as a convex combination of the vertices (extreme point) x_1, x_2, x_3 of the triangle.

Solution:

Let ABC be the triangle with vertices x_1, x_2, x_3 respectively and P be any point w inside the triangle. Join the points A and P and extend this line to meet the base BC at D (u). Since D is a point on the line segment BC, so u can be written as a convex combination of x_2 and x_3.

Thus $u = \lambda_1 x_2 + (1 - \lambda_1) x_3, 0 \le \lambda_1 \le 1.$ ...(1)

Again P is a point on the line segment AD, therefore

$$w = \lambda_2 x_1 + (1 - \lambda_2) u, 0 \le \lambda_2 \le 1$$
$$= \lambda_2 x_1 + (1 - \lambda_2) [\lambda_1 x_2 + (1 - \lambda_1) x_3], \text{ using (1)}$$
$$= \lambda_2 x_1 + \lambda_1 (1 - \lambda_2) x_2 + (1 - \lambda_1) (1 - \lambda_2) x_3$$

or $$w = \mu_1 x_1 + \mu_2 x_2 + \mu_3 x_3 \quad ...(2)$$

where $\mu_1 = \lambda_2, \mu_2 = \lambda_1 (1 - \lambda_2),$

$\mu_3 = (1 - \lambda_1) (1 - \lambda_2).$

Each of the μ_1, μ_2, μ_3 lies between 0 and 1 *i.e.*, $0 \le \mu_i \le 1$, i = 1, 2, 3 as $0 \le \lambda_1$ 1 and $0 \le \lambda_2 \le 1$.

Also $\mu_1 + \mu_2 + \mu_3 = \lambda_2 + \lambda_1 (1 - \lambda_2) + (1 - \lambda_1) (1 - \lambda_2)$

$$= \lambda_2 + \lambda_1 - \lambda_1 \lambda_2 + 1 - \lambda_1 - \lambda_2 + \lambda_1 \lambda_2 = 1$$

$\Rightarrow \mu_1 x_1 + \mu_2 x_2 + \mu_3 x_3$ is a convex combination of the points x_1, x_2, x_3.

Hence (2) is the required convex combination for the point w.

Example 5(a):

Sketch the convex polygon spanned by the following points in a two dimensional. Euclidean space. Which of these points are vertices> Express the other as the convex linear combination of the vertices

$$(0, 0), (0, 1), (1, 0), \left(\frac{1}{2}, \frac{1}{4}\right).$$

Solution:

The convex combinations of the points (0, 0), (1, 0); (0, 0), (0, 1) and (1, 0), (0, 1) give the line segments OA, OB and AB respectively. Thus the convex combination of points (0, 0), (1, 0) and (0, 1) is the interior of the triangle OAB.

The points O(0, 0), A(1, 0) and B(0, 1) are the vertices and the point $C\left(\frac{1}{2}, \frac{1}{4}\right)$ is the interior point of the convex polygon spanned by the given points.

To express $\left(\frac{1}{2}, \frac{1}{4}\right)$ as the linear combination of (0, 0), (0, 1), (1, 0).

Let $\left(\frac{1}{2}, \frac{1}{4}\right) = \lambda_1 (0, 0) + \lambda_2 (0, 1) + \lambda_3 (1, 0)$ where $\lambda_1 + \lambda_2 + \lambda_3 = 1$ and $\lambda_1, \lambda_2, \lambda_3 \ge 0$.

$\therefore \left(\frac{1}{2}, \frac{1}{4}\right) = (\lambda_3, \lambda 2)$ which gives $\lambda 2 = \frac{1}{4}, \lambda_3 = \frac{1}{2}$.

$\therefore \lambda_1 = 1 - \lambda_2 - \lambda_3 = 1 - \frac{1}{4} - \frac{1}{2} = \frac{1}{4}$.

Thus $\left(\frac{1}{2}, \frac{1}{4}\right) = \frac{1}{4}(0,0) + \frac{1}{4}(0,1) + \frac{1}{4}(1,0)$.

Example 5(b):

For any points x, y ∈ R^n, show that the line segment [x : y] is a convex set.

Solution:

Let $\quad u, v \in [x : y]$. Then

$u = \lambda_1 x + (1 - \lambda_1) y, 0 \le \lambda_1 \le 1$

and $\quad v = \lambda_2 x + (1 - \lambda_2) y, 0 \le \lambda_2 \le 1$.

Now let w be a point on the line segment joining the points u and v.

Then $w = \lambda u + (1 - \lambda) v, 0 \le \lambda \le 1$.

$= \lambda [\lambda_1 x + (1 - \lambda_1) y] + (1 - \lambda) [\lambda_2 x + (1 - \lambda_2) y]$

$= [\lambda\lambda_1 + (1 - \lambda) \lambda_2] x + [\lambda (1 - \lambda_1) + (1 - \lambda) (1 - \lambda_2)] y$.

If we take $\mu = \lambda\lambda_1 + (1 - \lambda) \lambda_2$, then

$1 - \mu = 1 - \lambda\lambda_1 - (1 - \lambda) \lambda_2 = \lambda + (1 - \lambda) - \lambda\lambda_1 - (1 - \lambda) \lambda_2$

$= \lambda (1 - \lambda_1) + (1 - \lambda) (1 - \lambda_2)$.

Since, $0 \le \lambda_1 \le 1, 0 \le \lambda_2 \le 1$

$\Rightarrow 0 \le \lambda\lambda_1 + (1 - \lambda) \lambda_2 \le 1$,

$w = \mu x + (1 - \mu) y, 0 \le \mu \le 1$

$\therefore w \in [x : y]$.

Hence the set [x : y] is a convex set.

Example 6(a):

Let A be an m × n matrix and b an m-vector, then show that {x ∈ R^n : Ax ≤ b} is a convex set.

Solution:

Let $A = [a_{ij}]_{m \times n}$, $x = (x_1, x_2, \ldots, x_n)$ and $b = (b_1, b_2, \ldots, b_m)$, then the set $S = \{x \in R^n : Ax \le b\}$ can be written in m-inequalities:

$a_{11} x_1 + a_{12} x_2 + \ldots + a_{1n} x_n \le b_1$

$a_{21} x_1 + a_{22} x_2 + ... + a_{2n} x_n \le b_2$

$\cdots \quad \cdots \quad \cdots \quad \cdots \quad \cdots$

$a_{m1} x_1 + a_{m2} x_2 + ... + a_{mn} x_n \le b_n.$

Thus, the set S is the intersection of m half spaces

$H_i = \{(x_1, x_2, ... ,x_n) : a_{i1} x_1 + a_{i2} x_2 + ... + a_{in} x_n \le b_i, i = 1, 2, ...,m\}.$

If follows that $S = \bigcap_{i=1}^{m} H_i$ is convex as each half space is convex.

Example 6(b):

Show that the set $S = \{x : x = (x_1, x_2, x_3), x_1^2 + x_2^2 + x_3^2 \le 1\}$ is a convex set.

Solution:

Let x, y ∈ S, where $x = (x_1, x_2, x_3)$, $y = (y_1, y_2, y_3)$.

Then, by the given condition, we have

$$x_1^2 + x_2^2 + x_3^2 \le 1 \qquad ...(1)$$

and $$y_1^2 + y_2^2 + y_3^2 \le 1. \qquad ...(2)$$

If $z = (z_1, z_2, z_3)$ is a point on the line segment joining the points x and y, then

$z = \lambda x + (1 - \lambda) y, 0 \le \lambda \le 1$

$\Rightarrow (z_1, z_2, z_3) = l (x_1, x_2, x_3) + (1 - \lambda) (y_1, y_2, y_3)$

$\Rightarrow z_1 = \lambda x_1 + (1 - \lambda) y_1, z_2 = \lambda x_2 + (1 - \lambda) y_2, z_3 = \lambda x_3 + (1 - \lambda) y_3.$

Now $z_1^2 + z_2^2 + z_3^2 = [\lambda x_1 + (1 - \lambda) y_1]^2 + [\lambda x_2 + (1 - \lambda) y_2]^2 + \lambda x_3 + (1 - \lambda) y_3]^2$

$= \lambda_2 (x_1^2 + x_2^2 + x_3^2) + (1 - \lambda)^2 (y_1^2 + y_2^2 + y_3^2) + 2\lambda (1 - \lambda) (x_1 y_1 + x_2 y_2 + x_3 y_3)$

$\le \lambda_2 . 1 + (1 - \lambda)^2 . \lambda + 2_l (1 - \lambda) (x_1 y_1 + x_2 y_2 + x_3 y_3)$, using (1) and (2)

By Lagrange's identity, we have

$(x_1^2 + x_2^2 + x_3^2) (y_1^2 + y_2^2 + y_3^2) - x_1 y_1 + x_2 y_2 + x_3 y_3)^2 = \Sigma (x_1 y_2 - x_2 y_1)^2 \ge 0$

$\therefore x_1 y_1 + x_2 y_2 + x_3 y_3 \le \sqrt{}(x_1^2 + x_2^2 + x_3^2) \sqrt{}(y_1^2 + y_2^2 + y_3^2) \le 1$, using (1) and (2)

Thus $z_1^2 + z_2^2 + z_3^2 \le \lambda_2 + \lambda_2 + (1 - \lambda)^2 + 2_l . (1 - \lambda) = 1.$

$\therefore z = (z_1, z_2, z_3) \in S.$

Hence the set S is a convex set.

Example 6(c):

Examine convexity to the set

$\{(x_1, x_2) \in R^2 : 4x_1 + 3x_2 \leq 6, x_1 + x_2 \geq 1\}$

Solution:

Let $S = \{(x_1, x_2) \in R^2 : 4x_1 + 3x_2 \leq 6$ and $x_1 + x_2 \geq 1$

If $u = (x_1, x_2) \in S$ Then $4x_1 + 3x_2 \leq 6$ and $x_1 + x_2 \geq 1$

and if $v = (y_1, y_2) \in S$ then $4y_1 + 3y_2 \geq 6$ and $x_1 + y_2 \geq 1$.

If $w = (z_1, z_2)$ is a point on the line segment joining points u and v, then

$$w = \lambda u + (1 - \lambda) v \qquad 0 \leq 1 \leq 1$$

$\Rightarrow (z_1, z_2) = \lambda (x_1, x_2) + (1 - \lambda) (y_1, y_2)$

$= (\lambda x_1 + (1 - \lambda) y_1, \lambda x_2 + (1 - \lambda) y_2)$

$\Rightarrow z_1 = \lambda x_1 + (1 - \lambda) y_1$ and $z_2 = \lambda x_2 + (1 - \lambda) y_2$

Now $4z_1 + 3z_2 = \lambda (4x_1 + 3x_2) + (1 - \lambda) (4y_1 + 3y_2) \leq 6^*$

$\because 0 \leq \lambda \leq 1, 0 \leq 1 - \lambda \leq 1, 4x_1 + 3x_2 \leq 6$ and $4y_1 + 3y_2 \leq 6$.

and $\quad z_1 + z_2 = \lambda (x_1 + x_2) + (1 - \lambda) (y_1 + y_2) \geq 1^{**}$

$\because 0 \leq \lambda \leq 1, 0 \leq 1 - \lambda \leq 1, x_1 + x_2 \geq 1, y_1 + y, \geq 1.$

$\because w = (z_1, z_2) \in S.$

Hence the set S is a convex set.

Note: As the extreme case if $4x_1 + 3x_2 = 6, 4y_1 + 3y_2 = 6$,

then $4z_1 + 3z_2 = \lambda. 6 + (1 - \lambda) 6 = 6$

and if $x_1 + x_2 = 1, y_1 + y_2 = 1$

then $z_1 + z_2 = \lambda. 1 + (1 - \lambda). 1 = 1$.

Aliter: Obviously S is the intersection of two half spaces, viz.,

$$H_1 = \{(x_1, x_2) : 4x_1 + 3x_2 \leq 6\}$$

and $\quad H_2 = \{(x_1, x_2) : x_1 + x_2 \geq 1\}$.

Since H_1 and H_2 are convex sets so $S = H_1 \cap H_2$ is also convex.

EXERCISES

1. (a) Prove that the collection of feasible solutions of a L.P.P. constitutes a convex set, whose extreme points correspond to B.F. solutions.

(b) Find an optimal solution of the following L.P.P. without using the simplex algorithm:

Max. $Z = 2x_1 + 3x_2 + 4x_3 + 7x_4$

such that $2x_1 + 3x_2 - x_3 + 4x_4 = 8$

$x_1 - 2x_2 + 6x_3 - 7x_4 = -3,$

$x_i \geq 0,\ i = 1, 2, 3, 4.$

2. If a set of $k \leq m$ vectors $P_1, P_2, \ldots, P_k \in A$ can be found that are linearly independent and such that $x_1 P_1 + x_2 P_2 + \ldots + x_k P_k = b$, and all $x_i \geq 0$, then show that the point $x = (x_1, x_2, \ldots x_k, 0, 0 \ldots, 0)$ is an extreme point of the convex set of feasible solution of $Ax = b$.
3. (a) If $x_1, x_2 \in S \Rightarrow 1/2\,(x_1 + x_2) \in S$ than the set S is convex or not.

   (b) Can there be any convex set without any extreme point? Prove that an extreme point of a convex set is a boundary point of the set.

   (c) Is the set $S = \{x : x \in E^m, |x| = 1\}$ convex?
4. Which of the following sets are convex?

   (i) $A = \{(x_1, x_2) : x_1 x_2 \leq 1, x_1 \geq 0, x_2 \geq 0\}$

   (ii) $A = \{(x_1, x_2) : x_2^2 - 3 \geq x_1^2, x_1 \geq 0, x_2 \geq 0\}$.
5. (a) Prove that the set $\{(x_1, x_2) : x_1^2 + x_2^2 \leq 4\}$ is a convex set.

   (b) Define a convex set. Show that the set $S = \{(x_1, x_2) : 3x_1^2 + 2x_2^2 \leq 6\}$ is convex.
6. Explain the procedure of generating extreme point solutions (Analytical method) to a linear programming problem pointing out the assumptions made, if any.
7. Given two planes $a_1x + b_1y + c_1z + d_1 = 0$, $a_2x + b_2y + c_2z + d_2 = 0$ in R^3, prove that their intersection is a convex set but their union is not.
8. Examine convexity of the following sets:

   (i) $S = \left\{(x_1, x_2) : \dfrac{x_1^2}{4} + \dfrac{x_2^2}{9} \leq 1\right\}$

   (ii) $S = \{(x_1, x_2) : x_1^2 + x_2^2 \leq 1, x_1 + x_2 \geq 1\}$
9. If S_1 and S_2 be two non-empty disjoint sets and S be a set such that if $x_1 \in S_1$ and $x_1 \in S_2$ then $x_1 - x_2 \in S$. Show that S is also a convex set and does not contain the origin.
10. Show that the set of all the internal points of a convex set S is a convex set.

11. Determine whether the vector [7, 0] is a convex combination of the vectors [6, 3], [9, – 6], [1, 2], [1, –1].
12. Determine the convex hull of the following sets:

    (i) $A = \{(x_1, x_2) : x_1^2 + x_2^2 = 1\}$ (ii) $A = \{x_1, x^2\}$.
13. A and B are two convex sets in R^n, C is a set in R^n defined as

    $C = \{z \in R^n : z = x + y, \; x \in A, y \in B\}$.

    Examine convexity of C.
14. Find the extreme points of the polygonal convex set X determined by the system

    $2x_1 + x_2 + 9 \geq 0, -x_1 + 3x_2 + 6 \geq 0,$

    $x_1 + x_2 \leq 0, x_2 + x_2 \leq 0, x_1 + 2x_2 - 3 \leq 0.$
15. Express (2, 1), $\left(0, \frac{3}{2}\right)$, if possible, as a convex combinations of (1, 1) and (–1, 2).
16. What is meant by convex combination of vectors? Prove that the set of all convex combinations of linearly independent vectors is a convex set.
17. Prove that the objective function of a L.P.P. assumes minimum value at an extreme point of the convex set X generated by the set of all feasible solutions.
18. Show that $S = \{(x_1, x_2, x_3) : 2x_1 - x_2 + x_3 \leq 4, x_1 + 2x_2 - x_3 \leq 1\}$ is a convex set.
19. Give an example of a convex set whose every boundary point is an extreme point.

MULTIPLE CHOICE QUESTIONS

1. A rectangle with sides a_1 and a_2 ($a_1 \neq a_2$) is placed with one corner at the origin and two of its sides along the axes. The interior of the rectangle plus its edges form a

    (a) convex set (b) non-convex set

    (c) polyhedron convex set (d) none of these
2. Which of the following sets in E2 is not a convex set:

    (a) $\{(x_1, x_2) : x_1^2 + x_2^2 \leq 1\}$

    (b) $\{(x_1, x_2) : x_1^2 + x_2^2 \leq 4\}$

    (c) $\{(x_1, x_2) : x_1^2 + x_2^2 \geq 1, x_1^2 + x_2^2 \leq 4\}$

    (d) $\{(x_1, x_2) : x_1 \geq 0\}$.

3. In a two dimensional Euclidean space the points (0, 0), (0, 1), (1, 0), $\left(\frac{1}{2}, \frac{1}{4}\right)$ span the convex polygon. Then the vertices of the polygon are

   (a) (0, 0), (1, 0), (0, 1) (b) (0, 0), (1, 0), $\left(\frac{1}{2}, \frac{1}{4}\right)$

   (c) (0, 0), (0, 1), $\left(\frac{1}{2}, \frac{1}{4}\right)$ (d) none of these.

4. The maximum number of extreme points for a L.P.P. max. Z = cx subject to Ax = b, x ≥ 0, where A is m × n matrix, is equal to

   (a) $\frac{m!}{n!(n-m)!}$ (b) $\frac{m!}{m!(n-m)!}$

   (c) m (d) n.

5. Consider the triangle with vertices (0, 0), (2, 0), (1, 1). The point (.3, 2.) as a convex combination of these vertices is

   (a) .75 (0, 0) + .05 (2, 0) + .2 (1, 1)

   (b) .25 (0, 0) + .50 (2, 0) + .25 (1, 1)

   (c) .30 (0, 0) + .60 (2, 0) + .10 (1, 1)

   (d) none of these.

6. The extreme points of the set $\{(x, y) : |x| \leq 1, |y| \leq 1\}$ are

   (a) (1, 1) (1, –1)

   (b) (1, 1), (1, –1), (–1, 1)

   (c) (1, 1), (1, –1), (–1, 1), (–1, –1)

   (d) (1, 1), (–1, –1).

7. The convex hull of the set of all the points on the boundary of the circle is the

   (a) interior of the circle (b) Whole circle

   (c) boundary of the circle (d) none of these.

3

Dual Simplex Method

INTRODUCTION

The simplex method is an algorithm that always deals with a basic feasible solution and the algorithm is terminated as soon as an optimal solution is achieved. That is, the procedure should be stopped when all $c_j - z_j \leq 0$ for maximization problem and $c_j - z_j \geq 0$ for minimization problem. However, if one or more solution values (i.e. x_{Bi}) are negative and optimality condition $c_j - z_j$ both for maximization and minimization is satisfied, then current optimal solution may not be feasible (because $c_j - z_j = c_j - c_B B^{-1} a_j$ is completely independent of the vector b). In such cases, it is possible to find a starting basic, but not feasible solution that is dual feasible, *i.e.*, all $c_j - z_j \leq 0$ for a miximization problem. In all such cases a variant of the simplex method called the *dual-simplex method* would be used. In the dual simplex method we always attempt to retain optimality while bringing the primal back to feasibility (*i.e.* $x_{Bi} \geq 0$ for all i).

DUAL-SIMPLEX ALGORITHM

The steps of a dual-simplex algorithm may be summarized as follows:

Step 1. Determine an Initial Solution

Convert the given LP problem into the standard form by adding slack, surplus and artificial variables and obtain an initial basic feasible solution. Display this solution in the initial dual-simplex table.

Step 2. Test Optimality of the Solution

If all solution values are positive (*i.e.*, $x_{Bi} \geq 0$ for all i), then there is no need of applying a dual-simplex method because improved solution can be obtained by simplex method itself. Otherwise go to Step 3.

Step 3. Test Feasibility of the Solution

If there exists a row, say r, for which solution value is negative (*i.e.*, $x_{Br} < 0$) and all elements in row r and column j are positive (*i.e.* $y_{rj} \geq 0$ for all j), then current solution is infeasible; hence go to Step 4.

Step 4. Revise the Solution

Repeat Steps 2 to 4 until either an optimal solution is reached or there exists no feasible solution. A flow chart of solution procedure of dual-simplex method is shown in Fig. 3.1.

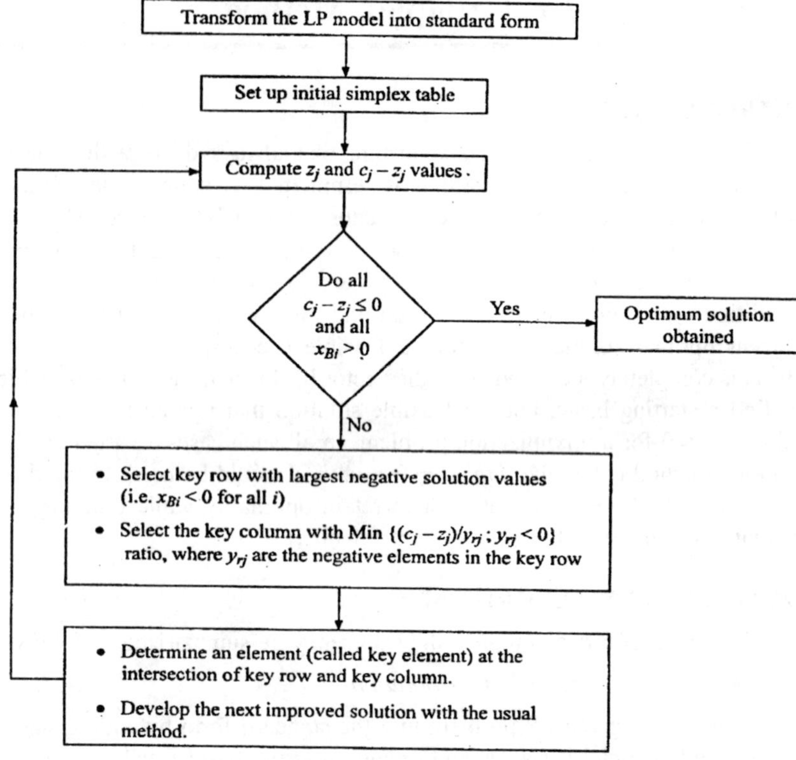

Fig. 3.1 : Flow Chart of Dual-Simplex Method.

Step 5. Obtain Improved Solution

(i) Select a basic variable associated with the row (called key row) having the largest negative solution value, *i.e.*

$$x_{Br} = \text{Min }\{x_{Bi};\ x_{Bi} < 0\}$$

(ii) Determine the minimum ratios only for those columns having a negative element in row r. Then select a non-basic variable for entering into the basis associated with the column for which

$$\frac{c_k - c_k}{y_{rk}} = \text{Min}\left\{\frac{c_j - z_j}{y_{rj}};\ y_{rj} < 0\right\}, \text{ for all } j$$

The element (*i.e.* y_{rk}) at the intersection of key row and key column is called *key element*. The improved solution can then be obtained by making y_{rk} as 1 and all other element of the key column zero. Here, it may be noted that key element is always positive.

DUAL SIMPLEX METHOD OF AN LINEAR PROGRAMMING PROBLEM WITH MIXED RESTRICTIONS

Sometimes a given L.P.P. contains a mixture of inequalities ($\geq$, $\leq$), equations; non-negative variables and unrestricted variables, then to obtain its dual we proceed in the following manner:

(i) If a constraint is an equation (has = sign), replace it by two constraints involving the inequalities going in opposite directions.

For example,

the equation $2x_1 + 5x_2 = 9$ is replaced by

$$2x_1 + 5x_2 \leq 9 \qquad ...(1)$$

and $$2x_1 + 5x_2 \geq 9. \qquad ...(2)$$

(ii) If the given problem is of maximization, all constraints should have $\leq$ sign. If some constraint has $\geq$ sign, multiply both sides by – 1 and make the sign $\leq$.

Thus in the above example multiply both sides of (2) by – 1 so that it becomes $-2x_1 - 5x_2 \leq -9$

Similarly, if the problem is of minimization, all constraints should have $\geq$ sign.

(iii) If there is some unrestricted variable, replace it by the difference of two variables.

Standard Primal Form

A linear programming problem is said to be in *standard primal form if*

(a) For a maximization problem all the constraints have $\leq$ sign.

(b) For a minimization problem all the constraints have $\geq$ sign.

MANAGERIAL SIGNIFICANCE OF PROBLEM ANALYSIS

The importance of the dual LP problem is in terms of the information which it provides about the value of the resources. The economic analysis is concerned in deciding whether or not to secure more resources and how much to pay for these additional resources.

The significance of the study of dual is as follows:

(i) The maximum amount that should be paid for one additional unit of a resource is called its shadow price (also called *simplex multiplier*).

(ii) The total marginal value of the resources equals the optimal objective function value. The dual variables equal the marginal value of resources (shadow prices).

(iii) The dual variables provide the decision-maker a basis for deciding how much to pay for additional units of resources.

The value of the ith dual variable is the rate at which the primal objective function value will increase as the ith resource increases, assuming that all other data is unchanged.

THE ASSIGNMENT PROBLEM (MAXIMAL)

The problem of minimizing the total cost. Sometimes the assignment problem deals with the maximization of an objective function rather than to minimize it. For example, the problem may be to assign persons to jobs in such a way that the expected profit is maximum. This problem may be solved easily by first converting it to a minimization problem and then we can use the usual procedure of assignment algorithm. This conversion can be very easily done by modifying the given profit matrix to the cost matrix in either of the two ways.

(i) *Select the greatest element of the given profit matrix and then subtract each element of the matrix from this greatest element to get the modified matrix.*

For, if $[c_{ij}]$ is the given profit matrix and c_{rk} is the greatest element of this matrix then the modified matrix will be $[c_{ij}']$, where $c_{ij}' = c_{rk} - c_{ij}$. It can be shown that if $x_{ij} = x_{ij}$ maximizes $Z = \sum\sum c_{ij}\, x_{ij}$, then $x_{ij} = X_{ij}$ minimizes the function $Z' = \sum\sum c_{ij}'\, x_{ij}$, where $c_{ij}' = c_{rk} - c_{ij}$. It follows from the relation

$$Z' = \sum\sum c_{ij}'\, x_{ij} = \sum\sum (c_{rk} - c_{ij})\, x_{ij} = \sum\sum c_{rk}\, x_{ij} - \sum\sum c_{ij}\, x_{ij} = nc_{rk} - Z$$

(ii) *Place minus sign before each element of the profit matrix to get the modified matrix.*

In this case if $[c_{ij}]$ is the given profit matrix then the modified matrix will be $[c_{ij}']$ where $c_{ij}' = -c_{ij}$. It can be shown that if $x_{ij} = X_{ij}$ maximizes $Z = \sum\sum c_{ij}\, x_{ij}$ then $x_{ij} = X_{ij}$ minimizes $Z' = \sum\sum c_{ij}'\, x_{ij}$.

USEFUL ASPECTS OF DUAL SIMPLEX PROBLEM

In some cases it is easier to solve linear programming problems through their duals. If the number of original variables in the primal problem is considerably less than the number of slack and surplus variables, we must opt to solve the problem through its dual.

Dual problem plays an important role not only in linear programming but also in physics, economics, engineering etc.

In physics, we use it in parallel circuit and series circuit theory.

In Economics it is used in the formulation of input and output systems.

In game theory, we use it to find the optimal strategies of the player B when he minimizes his losses. Then, using Dual problem, we can change the player A's problem into player B's problem and *vice-versa*.

COMPLEMENTARY SLACKNESS PRINCIPLE

The principle of complementary slackness establishes the relationship between the optimal value of a primal variable in one problem with a slack or surplus variable of the dual problem. To illustrate this concept, let us consider the example of production planning.

1. If at the optimal solution of a primal problem, a primal constraint has positive value of slack variable, the corresponding resource is not completely used up and must have zero opportunity cost (shadow price). This means that having more of this resource will not improve the value of the objective function. But if the value of slack variable is zero in that constraint, the entire resource is being used and must have a positive opportunity cost.

   Slack variables in a primal LP problem correspond to the dual variables in the dual LP problem. Thus, complementary slackness states that for every resource, the following condition must hold

   $$\text{Primal slack variable} \times \text{Dual variable} = 0.$$

2. When resources are not used completely for producing any unit of a product, in that case opportunity cost of such resources exceeds the unit profit from that product. But if some of the product is made, the opportunity cost of its resources must equal its unit profit, so that surplus product cost is zero. A primal variable in primal LP problem represents quantity of product made and surplus variables in dual represent surplus product costs. In this case complementary slackness states that for every unit of the product, following condition must hold.

   $$\text{Primal variable} \times \text{Dual surplus variable} = 0.$$

ADVANTAGES OF ANALYSIS PROBLEM

1. It is advantageous to solve the dual of a primal having less number of constraints, because the number of constraints usually equals the number of iterations required to solve the problem.

2. It avoids the necessity for adding surplus or artificial variables and solves the problem quickly (the technique is known as the primal-dual method). In economics, duality is useful in the formulation of the input and output systems. It is also useful in physics, engineering, mathematics, etc.
3. The dual variables provide an important economic interpretation of the final solution of an LP problem.
4. It is quite useful when investigating changes in the parameters of an LP problem (the technique is known as the sensitivity analysis).
5. Duality is used to solve an LP problem by the simplex method in which the initial solution is infeasible (the technique is known as the dual simplex method).

SOME ANALYSIS PROBLEM THEOREMS

We observe that dual of a primal problem is also a linear programming problem therefore we can also construct the dual of a dual problem.

Here we shall prove a number of fundamental theorems to describe the relation between the primal and its dual. The relationship between the primal and dual is extremely useful in the development of mathematical programming.

Theorem 1:

If the L.P.P. mini. $Z_p = c\,x$ *& &* $A\,v \geq b$ *has finites optimal solution than the dual max.* $Z = b'\,w$ *&* $A'\,w \leq c'$ *also has a finite maxi solution and if dual has a finite optimal solution then so as primal.*

Proof:

Let the primal is

Mini Z = c x

$$a_{11}\,x_1 + a_{12}\,x_2 + \dots + a_{1n}\,x_n \geq b_1$$

$$a_{m1}\,x_1 + a_{m2}\,x_2 + \dots + a_{mn}\,x_n \geq b_m$$

changing in equation

$$a_{11}\,x_1 + a_{12}\,x_2 \dots + a_{1n}\,x_n - x_{1s} = b_1$$

$$a_{m1}\,x_1 + a_{m2}\,x_2 + \dots + a_{mn}\,x_n - x_{ms} = b_m$$

The L.P.P. can be written as

$$A\,x - I\,X_s = b$$

Let x_B is a basic feasible solution which is optimal for the primal corresponding to base $B = (\alpha_1, \alpha_2, \dots \alpha_m)$

$\Rightarrow x_B$ minimum solution of primal

$\Rightarrow$for this solution all $Z_j - c_j \leq 0$

$\Rightarrow c_B B^{-1} \alpha_j - c_j \leq 0$ for all j either original or for surplus variables

$\Rightarrow c_B B^{-1} \alpha_1 \leq c_1$

$c_B B^{-1} \alpha_2 \leq c_2$

$c_B B^{-1} \alpha_n \leq c_n$ *i.e.,* for original variable

$\Rightarrow (c_B b^{-1}) (\alpha, \alpha, \dots, \alpha_n \leq (c_1, c_2 - c_n)$

$\Rightarrow w_B' A \leq c$

Taking transport let $(c_B B^{-1} = w_B')$

$\Rightarrow A' w_B \leq c'$

$\Rightarrow w_B$ satisfy $A' w \leq c'$ *i.e.,* of dual

$\Rightarrow w_B$ is a dual solution.

I[st] we prove w_B is feasible solution IInd we prove w_B is optimum solution

$w_B' = c_B (B^{-1})$

Since $Z_J - c_j \leq 0$ for surplus variables also

$\Rightarrow c_B B^{-1} \alpha_{js} - 0 \leq 0$

$$\Rightarrow (w_{B1}, w_{B2}, \dots, w_{Bm}) \begin{pmatrix} 0 \\ 0 \\ 0 \\ 1 \\ -1 \\ 0 \end{pmatrix}_{J^{\text{the}} \text{place}} \leq 0$$

$\Rightarrow - w_{BJ} \leq 0$ for $j = 1, 2, -m$

$\Rightarrow w_{BJ} \geq 0 \; j = 1, 2, \dots, m$

$\Rightarrow w_B = (c_B B^{-1})$

is a feasible solution of dual.

Theorem 2:

The dual of a dual of the given primal is the primal itself.

Proof:

Suppose the given linear programming problem is

Primal Problem

$$\text{Max. } Z_P = c_1 x_1 + c_2 x_2 + \ldots + c_n x_n$$

subject to

$$\begin{aligned} &a_{11} x_1 + a_{12} x_2 + \ldots + a_{1n} x_n \le b_1 \\ &a_{21} x_1 + a_{22} x_2 + \ldots + a_{2n} x_n \le b_2 \\ &\ldots \qquad \ldots \qquad \ldots \\ &\ldots \qquad \ldots \qquad \ldots \\ &a_{m1} x_1 + a_{m2} x_2 + \ldots + a_{mn} x_n \le b_m, \\ &x_1, x_2, \ldots , m_n \ge 0. \end{aligned} \qquad \ldots(1)$$

Dual Problem

The dual of the above primal can be written as

$$\text{Mini. } Z_D = b_1 w_1 + b_2 w_2 + \ldots + b_m w_m \ge c_1$$

subject to

$$\begin{aligned} &a_{11} w_1 + a_{21} w_2 + \ldots + a_{m1} w_m \ge c_1 \\ &a_{12} w_1 + a_{22} w_2 + \ldots + a_{m2} w_m \; {}^3 \; c_2 \\ &\ldots \qquad \ldots \qquad \ldots \\ &\ldots \qquad \ldots \qquad \ldots \\ &a_{1n} w_1 + a_{2n} w_2 + \ldots + a_{mn} w_m \; {}^3 \; c_n \\ &w_1, w_2, \ldots , w_m \ge 0. \end{aligned} \qquad \ldots(2)$$

Now we have to construct the dual of the above dual. First we shall change the above dual to standard maximization form.

$$\text{Max. } (-Z_D) = -b_1 w_1 - b_2 w_2 - \ldots - b_m w_m$$

subject to

$$\begin{aligned} &-a_{11} w_1 - a_{21} w_2 - \ldots - a_{m1} w_m \le -c_1 \\ &-a_{12} w_1 - a_{22} w_2 - \ldots - a_{m2} w_m \le -c_2 \\ &\ldots \qquad \ldots \qquad \ldots \\ &\ldots \qquad \ldots \qquad \ldots \\ &-a_{1n} w_1 - a_{2n} w_2 - \ldots - a_{mn} w_m \le -c_n \\ &w_1, w_2, \ldots , w_m \ge 0. \end{aligned} \qquad \ldots(3)$$

Dual of the Dual

Considering the above dual as primal, its dual can be written as

$$\text{Mini. } Z_y = -c_1 y_1 - c_2 y_2 - \dots - c_n y_n$$

subject to

$$\begin{aligned} -a_{11} y_1 - a_{12} y_2 - \dots - a_{1n} y_n &\geq -b_1 \\ -a_{21} y_1 - a_{22} y_2 - \dots - a_{2n} y_n &\geq -b_2 \\ \dots \quad \dots \quad \dots \\ \dots \quad \dots \quad \dots \\ -a_{m1} y_1 - a_{m2} y_2 - \dots - a_{mn} y_n &\geq -b_m, \\ y_1, y_2, \dots, y_n &\geq 0. \end{aligned} \qquad \dots(4)$$

Changing the above problem to maximization and multiplying each constraint by – 1, we get

$$\text{Max. } Z_y' = c_1 y_1 + c_2 y_2 + \dots + c_n y_n, \; (-Z_y = Z_y' \text{ say})$$

subject to

$$\begin{aligned} a_{11} y_1 + a_{12} y_2 + \dots + a_{1n} y_n &\leq b_1 \\ a_{21} y_1 + a_{22} y_2 + \dots + a_{2n} y_n &\leq b_2 \\ \dots \quad \dots \quad \dots \\ \dots \quad \dots \quad \dots \\ a_{m1} y_1 + a_{m2} y_2 + \dots + a_{mn} y_n &\leq b_m, \\ y_1, y_2, \dots, y_n &\geq 0. \end{aligned} \qquad \dots(5)$$

Which is identical to the given linear programming problem (primal problem).

Hence, the dual of the dual is the primal.

Theorem 3:

Prove that dual of a dual is primal problem.

Proof:

Let primal problem is

$$\begin{aligned} &\text{Max. } Z = c\,x \\ &\text{s.t.} \quad A\,x \leq b \\ &\qquad\quad x \geq 0 \end{aligned} \qquad \dots(1)$$

Dual is Mini. $Z_{D1} = b'\,w$

s.t. $A' w \geq c'$ ...(2)

and $w \geq 0$

Thus dual (2) can be written as

Max $(-Z_{D1}) = -b' w$

$= d w$ [Put $-b' = d$]

s.t. $-A' w \leq -c'$ ...(3)

$w \geq 0$

Dual of the 3rd constraint is

Mini $(-Z_{D1}) = (-c')' y$

s.t. $(-A')' y \geq (-b')'$ ...(4)

$y \geq 0$

The dual (4) can be written as

Max. $Z_{D2} = c y$

s.t. $A y \leq b$

$y \geq 0$

which is similar to (1) *i.e.*, the primal.

Theorem 4:

If Max. $Z_P = c x$

s.t. $A x \leq b$ *is primal*

$x \geq 0$

Mini. $Z_D = b' w$

s.t. $A' w \geq c'$ *is dual*

$w \geq 0$

Prove that $D_P \leq Z_D$ $\forall$ *x is feasible solution of primal.*

$\forall$ *w is feasible solution of dual.*

Proof:

Primal Max. $Z_P = c x$

s.t. $A x \leq b$ ...(1)

$x \geq 0$

Dual Mini. $Z_D = b' w$

s.t. $A' w \geq c'$ ...(2)

$w \geq$

x and w be any feasible solution to the primal (1) and dual problem (2) respectively.

If we multiply the w' a row matrix in the constraint of primal

$$w' A x \leq w' b \quad ...(3)$$

and multiply the x' a column matrix in the constraint of dual

$$x' A' w \leq x' c' \quad ...(4)$$

equation (4) can be written as

$$(A x)' w \geq (c x)'$$

Taking transpose both side

$[(A x)' w]' A' x' = (x A)' [(c x)]'$ Matrix problem $A' x' = (x A)'$

$$(A x)'' = A x$$

$$w' (A x)'' \geq (c x)''$$

$$w' A x \geq c x \quad ...(5)$$

from (1) and (5)

$$w' A x \geq Z_P \quad ...(6)$$

from (3) and (6)

$$w' b \geq w' A x \geq Z_P$$

$$(b' w)' \geq Z_P$$

(Since $(b' w)' = b' w$, as $b' w$ is a scalar value, not a matrix)

So $\quad b' w \geq Z_P \quad ...(7)$

from (2) and (7)

$$Z_D \geq Z_P.$$

Theorem 5:

If $\overline{x}$ and $\overline{w}$ are feasible solution of primal & usual respectively such that $c\,\overline{x} = b'\,\overline{w}$ (i.e., $Z_P = Z_D$) then x and w are optimum solution of the respective problem respectively i.e., $\overline{x}$ is max and $\overline{w}$ is mini problem

Max $\quad Z_P = c x$

s.t. $\quad A x \leq b,\ x \geq 0$

and its corresponding dual

Mini. $Z_D = b' w$

$A' w \geq c'\ w \geq 0.$

Or Alternative Statement

The necessary and sufficient condition for any L.P.P. and its dual to have optimal solution is that both have feasible solution.

Proof:

Let $\overline{x}$ is any feasible solution to the primal

Max. $Z_P = c\,\overline{x}$, ...(1)

s.t. $A\,x \leq b,\ x \geq 0$

and w is any feasible solution of dual

Mini. $Z_D = b'\,w$

s.t. $A'\,w \geq c',\ w \geq 0$ ...(2)

then from theorem (2), we have

$c\,\overline{x} \leq b'\,w$ $(Z_P \leq Z_D)$

$b'\,\overline{w} = c\,x \leq b'\,w$ (as $c\,\overline{x} = b'\,w$ given)

i.e., $b'\,\overline{w} \leq b'\,w$

$\Rightarrow \overline{w}$ is mini solution of dual let $\overline{w}$ is given feasible solution of dual x is any feasible solution of primal

$c\,x \leq b'\,\overline{w} = c\,\overline{x}$

$\Rightarrow$ $c\,x \leq c\,\overline{x}$

$\Rightarrow$ $c\,\overline{x}$ is max value of the objective function

$\Rightarrow$ $\overline{x}$ is maximum solution of the primal problem.

Theorem 6:

If any of the constraints in the primal is a strict equality, the corresponding dual variable is unrestricted in sign.

Proof:

Let L.P.P. is

Mini $Z_P = c\,x$

s.t. $\sum_{j=1}^{n} a_{ij}x_j \geq b_i\ \forall_i$

and $\sum_{j=1}^{n} a_{kj}x_j = b_k\ \forall_i \geq k$

$x_j \geq 0 \ \forall_j$

In the standard primal form the above L.P.P. can be written as

Min $Z_P = c\,x$

s.t. $\sum_{j=1}^{n} a_{ij}x_j \geq b_i \ \forall_i \geq k$

and $\sum_{j=1}^{n} a_{kj}x_j \geq b_k$

$-\sum_{j=1}^{n} a_{kj}x_j \geq -b_k$

Now the dual of the above problem becomes

Max. $Z_D = \sum_{i=1}^{m} b_i w_i + b_k (w_k' - w_k'')$

$w_i \geq 0,\ i \neq k,\ w_k',\ w_k'' \geq 0$

Now writing $w_k'' - w_k'' = w_k$, the dual problem becomes

Max. $Z_D = \sum_{i=1}^{m} b_i w_i$

s.t. $\sum_{i=1}^{m} a_{ji} w_j \leq c_j \ \forall_j = 1, 2, \ldots n$

$w_j \geq 0 \ \forall_j \geq k$

obviously w_k is unrestricted in sign since $w_k >$ or < 0 according as $w_k' >$ or w_k''.

Theorem 7:

If x is any feasible solution to the primal problem Max $Z_P = c\,x$, subject to $A\,x \leq b,\ x \geq 0$ and w is any feasible solution to the dual problem

Min $Z_D = b'\,w$

subject to $A'\,w \geq c',\ w \geq 0$,

then $c\,x \leq b'\,w$ i.e., $Z_P \leq Z_D$.

Proof:

Consider the primal problem

Max. $Z_P = c\,x$

subject to $A\,x \leq b$ ...(1)

$x \geq 0.$

Let $x = (x_1, x_2, \dots, x_n)$ be any feasible solution to (1).

The dual of the above primal is

$$\text{Min. } Z_D = b' w$$

subject to $A' w \geq c'$, $w \geq 0$. ...(2)

Let $w = (w_1, w_2, \dots w_m)$ be any feasible solution to the dual (2).

Now w (m-component column vector) is the feasible solution of the dual and $A x \leq b$ are the constraints of the primal (1). If we multiply both sides of $A x \leq b$ with w', the sign of the inequality remains unchanged.

$$\therefore \quad w' (A x) \leq w' b$$

$$\Rightarrow \quad (A' w)' x \leq (b' w)', \qquad \text{...(3)}$$

Similarly, x (n-component column vector) is the feasible solution of the primal (1) and $A'w \geq c'$ denotes the constraints of the dual (2) so we can write

$$x' (A' w) \geq x' c'$$

$$\Rightarrow \quad x' (w' A)' \geq (c x)'$$

$$\Rightarrow \quad [(w' A)x]' \geq (c x)'$$

$$\Rightarrow \quad (w' A) x \geq c x$$

$$\Rightarrow \quad (A' w)' x \geq c x. \qquad \text{...(4)}$$

From relations (3) and (4) we have

$$c x \geq (A' w)' x \leq (b' w)'$$

$$\Rightarrow \quad c x \leq (b' w)' \Rightarrow c x \leq b' w$$

$$\Rightarrow \quad Z_P \leq Z_D.$$

Theorem 8:

If $\hat{x}$ is feasible solution to the primal problem Max. $Z_P = c x$ subject to $A x \leq b$, $x \geq 0$ and $\hat{w}$ is a feasible solution to its dual

$$\text{Min. } Z_D = b' w \text{ subject to } A' w \geq c', \; w \geq 0$$

such that $c \hat{x} = b' \hat{w}$, then x is the optimal solution of the primal and $\hat{w}$ is the optimal solution of the dual problem. or

The necessary and sufficient condition for any linear programming problem and its dual to have optimal solution is that both have feasible solution.

Proof:

Consider the primal problem

$$\text{Max. } Z_P = c\,x$$

subject to $A\,x \le b,\ x \ge 0.$ ...(1)

Let x be any feasible solution to (1).

The dual of the problem (1) is

$$\text{Min. } Z_D = b'\,w \qquad ...(2)$$

subject to $A'\,w \ge c',\ w \ge 0$

Let $\hat{w}$ be any feasible solution to the dual (2).

Proceeding as in theorem (2) above, we have

$$c\,x \le b'\,\hat{w}$$

$$\Rightarrow \qquad c\,x \le c\,\hat{x},\ \text{since } c\,\hat{x} = b'\,\hat{w}$$

Value of the objective function of the primal problem at the feasible solution $\hat{x}$ is greater than its value at any other feasible solution $\hat{x}$.

Hence $\hat{x}$ is the optimal solution of the primal problem (for maximization problem).

Again suppose w is any feasible solution of the dual problem (2). x is the given feasible solution of the primal problem (1),

Proceeding as in theorem (2), we have

$$c\,x \le b'\,w$$

$$\Rightarrow \qquad b'\,\hat{w} \le b'\,w \text{ since } c\,x = b'\,\hat{w}$$

The value of the objective function of the dual problem at the given feasible solution $\hat{w}$ is less than its value at any other feasible solution $\hat{w}$.

Hence $\hat{w}$ is the optimal solution of the dual problem (for minimization problem).

Theorem 9:

If any variable of the primal is unrestricted in sign, the corresponding constraint in the dual will be a stick equality.

Proof:

Let the kth variable x_k of the primal be unrestricted in sign.

Therefore replacing x_k by $x_k' - x_k''$ where x_k'' and x_k'' are non-negative variables, the primal can be written as

Max. $Z_D = c_1 x_1 + c_2 x_2 + \ldots + c_k (x_k' - x_k'') + c_{k+1} x_{k+1} + \ldots + c_n x_n$

s.t. $a_{11} x_1 + a_{12} x_2 + \ldots + a_{1k} (x_k' - x_k'') + a_{k+1} x_{k+1} + \ldots + a_{1n} x_n \le b_1$

....

....

....

$a_{m1} x_1 + a_{m2} x_2 + ... + a_{mk} (x_k' - x_k'') + a_{m}, b_{11} x_{b+1} + a_{mn} x_n \leq b_m$

and $\quad x_1, ... , x_{k-1}, x_k', x_b'', x_{k+1}, ... , x_n \geq 0$

The dual of the above problem is given by

Min. $Z_D = b_1 w_1 + b_2 w_2 + ... + b_m w_m$

and $\quad a_{11} w_1 + a_{21} w_2 + ... + a_{m1} w_m \geq c_1$

....

....

$a_{12} w_1 + a_{22} w_2 + ... + a_{m2} w_m \geq c_2$

$a_{1k} w_1 + a_{2k} w_2 + ... + a_{mk} w_m \geq c_k$

....

....

$a_{1k} w_1 - a_{2k} w_2 - ... - a_{mk} w_m \geq - c_k$

$a_{1n} w_1 + a_{2n} w + ... + a_{mn} w_m \geq c_n$

and $\quad w_1, w_2, ... , w_m \geq 0$

The two constraints under bracket {} in the above dual are equivalent to an equality

$$a_{1k} w_1 + a_{2k} w_2 + ... + a_{mk} w_m = c_k$$

i.e., the kth constraint in the dual is an equality.

Theorem 10:

(i) If either the primal or the dual problem has a finite optimal solution then the other problem so has a finite optimal solution and the optimal values of the objective function in both the problems are the same.

(ii) If primal (dual) problem has an unbounded optimum solution, the other problem has either no solution at all or an unbounded solution.

Proof:

Consider the primal and the dual problems

Primal problem

Max. $Z_P = c\,x$

subject $A\,x \leq b, x \geq 0$ ...(i)

Dual problem.

Min. $Z_D = b' w$

subject to $A' w \geq c'$ ...(ii)

$w \geq 0$

To prove the theorem we shall construct an optimal solution to the dual form a given optimal solution of the primal.

Let us assume that the primal has a finite optimal feasible solution x_B.

To solve the primal (i) by simplex method, introduce slack variables to each of the constraints.

Now (i) can be written as

Max. $Z_P = c\,x$

subject $A\,x + Ix_s = b$ ...(iii)

$x \geq 0,\ x_s \geq 0,$

where $x_s \in R^m$ represents the vector of slack variables and I is the associated $m \times n$ identity matrix. B is the basis matrix and suppose c_B is the m-component row vector containing the prices of the basic variables.

Now x_B is the optimal solution to the primal so we have

$c_j - Z_j \leq 0,\ \forall_j.$

We know $Z_j = c_B\,Y_j = c_B\,B^{-1}\,\alpha_j$

$c_j - c_B\,B^{-1}\,\alpha_j \leq 0,\ \forall a_j$

or $\quad c_B\,B^{-1}\,\alpha_j \geq c_j$ ...(iv)

$\Rightarrow c_B\,B^{-1}\,(\alpha_1, \alpha_2, \ldots, \alpha_n) \geq (c_1, c_2, \ldots, c_n)$

$\Rightarrow c_B\,B^{-1}\,A \geq c$ ...(v)

Taking $(\hat{w})' = c_B\,B^{-1}$, where $\hat{w} = (w_1, w_2, \ldots, w_m)$.

Then from (v), we have

$(w)'\,A \geq c \Rightarrow [(w)'\,A]' \geq c'$

$\Rightarrow A'\,(\hat{w}) \geq c'$

$\Rightarrow \hat{w}$ is the solution of the dual (ii).

Again considering the relation (iv) with a_j corresponding to slack variables, we have

$c_B\,B^{-1}\,(e_j) \geq 0,\ j = 1, 2, \ldots, m;\ c_j = 0.$

$\Rightarrow (\hat{w})'\,e_j \geq 0 \Rightarrow (\hat{w})' \geq 0$

$\Rightarrow \hat{w} \geq 0$

$\Rightarrow$w is the feasible solution of the dual (ii).

Now it remains to show that w is an optimal solution to the dual (ii).

We have

$$Z_D = b'\,\hat{w} = [(\hat{w})'\,b]' = (\hat{w})'\,b$$
$$= (c_B\,B^{-1})\,b = c_B\,(B^{-1}\,b)$$
$$= c_B\,x_B = Z_P.$$

Since w and x_B are the feasible solution of the dual (ii) and the primal (1) respectively and

$$Z_D = Z_P$$

therefore $\hat{w}$ is the optimal solution of the dual (ii).

(ii) We shall prove it by contradiction. Suppose, when the primal problem has an unbounded solution, the dual problem has a finite optimal solution.

We have proved in theorem 1 that dual of the dual is the original primal. Also part (i) of this theorem we have proved that if a primal has a finite optimal solution, the dual also has finite optimal solution.

Thus, if we consider the dual as the primal, then its (which is the original primal) must have a finite optimal solution, which contradicts the hypothesis. Hence the dual has no finite optimal solution. If the primal problem has an unbounded solution, either the dual has no solution or an unbounded solution.

Similarly, we can prove that if the dual has an unbounded solution, the primal has no solution or an unbounded solution.

Theorem 11:

If any of the constraints in the primal is a perfect equality, the corresponding dual variable is unrestricted in sign.

Proof:

Suppose in the given primal the kth constraint is an equality. Writing the primal in the standard primal form, we have

$$\text{Max. } Z = c_1\,x_1 + c_2\,x_2 + \ldots + c_n\,x_n$$

subject to

$$a_{11}\,x_1 + a_{12}\,x_2 + \ldots + a_{1n}\,x_n \le b_1$$
$$a_{21}\,x_1 + a_{22}\,x_2 + \ldots + a_{2n}\,x_n \le b_2$$
$$\ldots \qquad \ldots \qquad \ldots$$
$$\ldots \qquad \ldots$$

$$a_{k1}\, x_1 + a_{k2}\, x_2 + \dots + a_{kn}\, x_n \le b_k$$
$$-\, a_{k1}\, x_1 - a_{k2}\, x_2 - \dots - a_{kn}\, x_n \le -\, b_k$$
$$\dots \qquad \dots \qquad \dots$$
$$\dots \qquad \dots \qquad \dots$$
$$a_{m1}\, x_1 + a_{m2}\, x_2 + \dots + a_{mn}\, x_n \le b_m,$$
$$x_1, x_2, \dots , x_n \ge 0.$$

The dual of the above primal is

$$\text{Min } Z_D = b_1\, w_1 + b_2\, w_2 + \dots + b_k\, (w_k' \quad w_k'') + \dots + b_m\, w_m$$

subject to

$$a_{11}\, w_1 + a_{21}\, w_2 + \dots + a_{k1}\, (w_k' - w_k'') + \dots + a_{m1}\, w_m \ge c_1$$
$$a_{12}\, w_1 + a_{22}\, w_2 + \dots + a_{k2}\, (w_k' - w_k'') + \dots + a_{m2}\, w_m \ge c_2$$
$$\dots \qquad \dots \qquad \dots$$
$$\dots \qquad \dots \qquad \dots$$
$$a_{1n}\, w_1 + a_{2n}\, w_2 + \dots + a_{kn}\, (w_k' - w_k'') + \dots + a_{mn}\, w_m \ge c_n$$
$$w_1, w_2, \dots , w_k', w_k'', \dots , w_m \ge 0.$$

Substituting $w_k = w_k' - w_k''$ in the above dual, we get

$$\text{Min. } Z_D = b_1\, w_1 + b_2\, w_2 + \dots + b_k\, w_k + \dots + b_m\, w_m$$

subject to

$$a_{11}\, w_1 + a_{21}\, w_2 + \dots + a_{k1}\, w_k + \dots + a_{m1}\, w_m \ge c_1$$
$$a_{12}\, w_1 + a_{22}\, w_2 + \dots + a_{k2}\, w_k + \dots + a_{m2}\, w_m \ge c_2$$
$$\dots \qquad \dots \qquad \dots$$
$$\dots \qquad \dots \qquad \dots$$
$$a_{1n}\, w_1 + a_{2n}\, w_2 + \dots + a_{kn}\, w_k + \dots + a_{mn}\, w_m \ge c_n$$
$$w_1, w_2, \dots , w_k - 1, w_k + 1, \dots , w_m \ge 0.$$

wk is unrestricted in sign because

$$w_k > 0 \text{ if } w_k' > w_k'' \text{ and } w_k < 0 \text{ if } w_k' < w_k''.$$

Theorem 12:

If any variable of the primal is unrestricted in sign, the corresponding constraint in the dual will be a strict equality.

Proof:

Consider the primal in which the kth variable is unrestricted in sign.

$$\text{Max. } Z = c_1 x_1 + c_2 x_2 + .. + c_k x_k + ... + c_n x_n$$

subject to

$$a_{11} x_1 + a_{12} x_2 + ... + a_{1k} x_k + ... + a_{1n} x_n \le b_1$$
$$a_{21} x_1 + a_{22} x_2 + ... + a_{2k} x_k + ... + a_{2n} x_n \le b_2$$
$$... \quad ... \quad ...$$
$$... \quad ... \quad ...$$
$$a_{m1} x_1 + a_{m2} x_2 + ... + a_{mk} x_k + ... + a_{mn} x_n \le b_m$$
$$x_1, x_2, ... , x_k - 1, x_{k+1} ... , x_n \ge 0,$$

x_k is unrestricted in sign.

Substituting $x_k = x_k' - x_k''$, $x_k' \ge 0$ in the given primal, it changes to

$$\text{Max. } Z = c_1 x_1 + c_2 x_2 + ... + c_k (x_k' - x_k'') + .. + c_n x_n$$

subject to

$$a_{11} x_1 + a_{12} x_2 + ... + a_{1k} (x_k' - x_k'') + ... + a_{1n} x_n \le b_1$$
$$a_{21} x_1 + a_{22} x_2 + ... + a_{2k} (x_k' - x_k'') + ... + a_{2n} x_n \le b_2$$
$$... \quad ... \quad ...$$
$$... \quad ... \quad ...$$
$$a_{m1} x_1 + a_{m2} x_2 + ... + a_{mk} (x_k' - x_k'') + ... + a_{mn} x_n \le b_m,$$
$$x_1, x_2, ... , x_{k-1}, x_k', x_k'', x_{k+1}, ... , x_n \ge 0.$$

Writing the dual of the above primal, we have

$$\text{Min. } Z_D = b_1 w_1 + b_2 w_2 + .. + b_m w_m,$$

subject to

$$a_{11} w_1 + a_{21} w_2 + ... + a_{m1} w_m \ge c_1$$
$$a_{12} w_1 + a_{22} w_2 + ... + a_{m2} w_m \ge c_2$$
$$... \quad ... \quad ...$$
$$... \quad ... \quad ...$$
$$a_{1k} w_1 + a_{2k} w_2 + ... + a_{mk} w_m \ge c_k$$
$$- a_{1k} w_1 - a_{2k} w_2 - ... - a_{mk} w_m \ge - c_k$$
$$... \quad ... \quad ...$$
$$... \quad ... \quad ...$$
$$a_{1n} w_1 + a_{2n} w_2 + ... + a_{mn} w_m \ge c_n,$$
$$w_1, w_2, ... , w_m \ge 0.$$

The two constraints

$$a_{1k} w_1 + a_{2k} w_2 + ... + a_{mk} w_m \geq c_k$$

and $$- a_{1k} w_1 - a_{2k} w_2 - ... - a_{mk} w_m \geq - c_k$$

are equivalent to the single equation

$$a_{1k} w_1 + a_{2k} w_2 + ... + a_{mk} w_m = c_k.$$

Hence if the kth variable of the primal is unrestricted, the kth constraint in the dual is an equality.

Theorem 13:

If there does not exist feasible solution to the dual (primal) but there exists at least one to the primal (dual), then there dies not exist any finite optimum solution to the primal (dual).

Proof:

Consider the primal problem

$$\text{Max. } Z = c\,x,$$

subject to $A\,x \leq b$, $x \geq 0$.

The corresponding dual problem is

$$\text{Min. } Z_D = b'\,w$$

subject to $A'\,w \geq c'$, $w \geq 0$.

Suppose there does not exist any feasible solution to the dual there does one to the primal, say $\hat{x}$. Then $c\,\hat{x}$ is the value of the primal objective function.

If we suppose that $\hat{x}$ is an optimum solution to the primal then by fundamental theorem of Dual problem there must exist a feasible solution to the dual which contradicts the hypothesis.

Therefore no feasible solution to the primal can be optimal.

Similarly, we can start with the primal.

Theorem 14:

There exist a bounded (finite) optimum solution to a linear programming problem if and only if there exists a feasible solution to both primal and its dual.

Proof:

Suppose Max. $Z = c\,x$

subject to $A\,x \leq b$, $x \geq 0$

and Min. $Z_D = b'\,w$

subject to $A' w \geq c'$, $w \geq 0$

are the primal and dual problems respectively.

Again suppose there exists an optimum feasible solution to the primal problem. Then by fundamental theorem of Dual problem the dual problem has atleast one feasible solution. Conversely, let us assume that both primal and the dual possess feasible solution x and w respectively. Then $c\hat{x}$ and $b'\hat{w}$ are both finite and $c\hat{x} \leq b'\hat{w}$ *i.e.*, $b'\hat{w}$ acts as an upper bound on $c\hat{x}$ although not necessarily the least upper bound. Hence the primal must have finite optimum solution.

SOLVED EXAMPLES

Example 1:

Write the dual of the following problem:

$$\text{Min. } Z = 2x_2 + 5x_3$$

$$\text{subject } x_1 + x_2 \geq 2$$

$$2x_1 + x_2 + 6x_3 \leq 6$$

$$x_1 - x_2 + 3x_3 = 4$$

$$x_1, x_2, x_3 \geq 0.$$

Solution:

First we shall convert the given problem to standard primal form:

1. Since the problem is of minimization therefore all the constraints should have the sign $\geq$.
2. Multiplying the second constraint by – 1, it becomes
   $-2x_1 - x_2 - 6x_3 \geq -6$.
3. Since the third constraint is an equality so replacing it by the following two constrains

$$x_1 - x_2 + 3x_3 \leq 4$$

$$x_1 - x_2 + 3x_3 \geq 4$$

or $$-x_1 + x_2 - 3x_3 \geq -4$$

and $$x_1 - x_2 + 3x_3 \geq 4.$$

$\therefore$ the given problem in standard primal form is

$$\text{Min. } Z = 0x_1 + 2x_2 + 5x_3$$

subject to $x_1 + x_2 \geq 2$

$$-2x_1 - x_2 - 6x_3 \geq -6$$
$$-x_1 + x_2 - 3x_3 \geq -4$$
$$x_1 - x_2 + 3x_3 \geq 4$$
$$x_1, x_2, x_3 \geq 0.$$

The matrix form of the above problem is

Min. $Z = (0, 2, 5)\,(x_1, x_2, x_3)$

subject to

$$\begin{bmatrix} 1 & 1 & 0 \\ -2 & -1 & -6 \\ -1 & 1 & -3 \\ 1 & -1 & 3 \end{bmatrix} \begin{bmatrix} x_1 \\ x_2 \\ x_3 \end{bmatrix} \geq \begin{bmatrix} 2 \\ -6 \\ -4 \\ 4 \end{bmatrix}$$

Now the dual of the given primal is

Max. $Z_D = (2, -6, -4, 4)\,(y_1, y_2, y_3', y_3'')$

subject to

$$\begin{bmatrix} 1 & -2 & -1 & 1 \\ 1 & -1 & 1 & -1 \\ 0 & -6 & -3 & 3 \end{bmatrix} \begin{bmatrix} y_1 \\ y_2 \\ y_3' \\ y_3'' \end{bmatrix} \leq \begin{bmatrix} 0 \\ 2 \\ 5 \end{bmatrix},$$

$$y_1, y_2, y_3', y_3'' \geq 0$$

or Max. $Z_D = 2y_1 - 6y_2 - 4\,(y_3' - y_3'')$

subject to

$$y_1 - 2y_2 - (y_3' - y_3'') \leq 0$$
$$y_1 - y_2 + y_3' - y_3'' \leq 2$$
$$-6y_2 - 3\,(y_3' - y_3'') \leq 5,$$
$$y_1, y_2, y_3', y_3'' \geq 0.$$

Substituting $y_3 = y_3' - y_3''$, the required dual is

Max. $Z_D = 2y_1 - 6y_2 - 4y_3$

subject to

$$y_1 - 2y_2 - y_3 \leq 0$$
$$y_1 - y_2 + y_3 \leq 2$$
$$-6y_2 - 3y_3 \leq 5,$$

$y_1, y_2 \geq 0$ and y_3 is unrestricted in sign.

Example 2:

Write the dual of the following problem

$$\text{Min. } Z = 10x_1 + 20x_2$$

subject to $3x_1 + 2x_2 \geq 18$

$x_1 + 3x_2 \geq 8$

$2x_1 - x_2 \leq 6$

$x_1, x_2 \geq 0.$

Solution:

First we shall convert the given L.P.P. into standard primal form.

1. Since the given problem is of minimization, therefore all the constraints should have $\geq$ sign.
2. The first two inequalities are in the right direction while the third one is not.
3. Multiplying both sides of the third inequality by -1 it changes to $-2x_1 + x_2 \geq -6$.
4. Thus the standard primal form of the given L.P.P. is

$$\text{Min } Z = 10x_1 + 20x_2$$

subject $3x_1 + 2x_2 \geq 18$

$x_1 + x_2 \geq 8$

$-2x_1 + x_2 \geq 6.$

and $x_1, x_2 \geq 0.$

5. The matrix form of the above L.P.P. is

$$\text{Min. } Z = (10, 20)(x_1, x_2)$$

$$= c\,x$$

subject

$$\begin{bmatrix} 3 & 2 \\ 1 & 3 \\ -2 & 1 \end{bmatrix} \begin{bmatrix} x_1 \\ x_2 \end{bmatrix} \geq \begin{bmatrix} 18 \\ 8 \\ -6 \end{bmatrix}$$

or $A\,x \geq b,$

$x_1, x_2 \geq 0.$

$\therefore$ the required dual is

Max. $Z_D = b'y = (18, 8, -6)(y_1, y_2, y_3)$

s.t. $A'y = c'$ or $\begin{bmatrix} 3 & 1 & -2 \\ 2 & 3 & 1 \end{bmatrix}\begin{bmatrix} y_1 \\ y_2 \\ y_3 \end{bmatrix} \le \begin{bmatrix} 10 \\ 20 \end{bmatrix}$

or Max. $Z_D = 18y_1 + 8y_2 - 6y_3$

s.t. $3y_1 + y_2 - 2y_3 \le 10$

$2y_1 + 3y_2 + y_3 \le 20,$

$y_1, y_2, y_3 \ge 0.$

Example 3:

Write the dual simplex method of the following problem:

Min. $Z = x_1 + x_2 + x_3$

subject to

$x_1 - 3x_2 + 4x_3 = 5,$

$x_1 - 2x_2 \le 3,\ 2x_2 - x_3 \ge 4,$

$x_1, x_2 \ge 0$, x_3 *is unrestricted.*

Solution:

First we shall write the given L.P.P. in the standard primal form, substituting $x_3 = x_3' - x''$, $x_3' \ge 0$, $x_3' \ge 0$. The given problem can be written in the standard primal form as

Min. $Z = x_1 + x_2 + x_3' - x_3$

subject to

$-x_1 + 3x_2 - 4(x_3' - x_3'') \ge -5$

$x_1 - 3x_2 + 4(x_3' - x_3'') \ge 5$

$-x_1 + 2x_2 \ge -3$

$2x_2 - (x_3' - x_3'') \ge 4,$

$x_1, x_2, x_3', x_3'' \ge 0.$

The matrix form of the above problem is

Min. $Z = (1, 1, 1, -1)(x_1, x_2, x_3', x_3'')$

subject to $\begin{bmatrix} -1 & 3 & -4 & 4 \\ 1 & -3 & 4 & -4 \\ -1 & 2 & 0 & 0 \\ 0 & 2 & -1 & 1 \end{bmatrix}\begin{bmatrix} x_1 \\ x_2 \\ x_3' \\ x_3'' \end{bmatrix} \ge \begin{bmatrix} -5 \\ 5 \\ -3 \\ 4 \end{bmatrix}$

Now the dual of the given primal is

$$\text{Max } Z_D = (-5, 5, -3, 4)\,(y_1', y_1'', y_2, y_3)$$
$$= -5\,(y_1' - y_1'') - 3y_2 + 4y_3$$

subject to
$$\begin{bmatrix} -1 & 1 & -1 & 0 \\ 3 & -3 & 2 & 2 \\ -4 & 4 & 0 & -1 \\ 4 & -4 & 0 & 1 \end{bmatrix} \begin{bmatrix} y_1' \\ y_1'' \\ y_2 \\ y_3 \end{bmatrix} \le \begin{bmatrix} 1 \\ 1 \\ 1 \\ -1 \end{bmatrix}$$

or
$$-(y_1' - y_1'') - y_2 \le 1$$
$$3\,(y_1' - y_1'') + 2y_2 + 2y_3 \le 1$$
$$-4\,(y_1' - y_1'') - y_3 \le 1$$
$$4\,(y_1' - y_1'') + y_3 \le -1,$$
$$y_1', y_1'', y_2, y_3 \ge 0.$$

Substituting $y_1 = y_1' - y_1''$, the required dual is

Max. $Z_D = -5y_1 - 3y_2 + 4y_3$

subject to $-y_1 - y_2 \le 1$

$3y_1 + 2y_2 + 2y_3 \le 1$

$-4y_1 - y_3 \le 1$

$4y_1 + y_3 \le -1$ or $-4y_1 - y_3 \ge 1,$

$y_2, y_3 \ge 0$ and y_1 is unrestricted.

Example 4:

Write the simplex dual of the L.P. problem

Max. $Z = 2x_1 + x_2$

s.t. $x_1 + 2x_2 \le 10$

$x_1 + x_2 \le 6$

$x_1 - x_2 \le 2$

$x_1 - 2x_2 \le 1$

$x_1, x_2, x_3 \ge 0.$

Solution:

In matrix from the given problem can be written as

Max. $Z = (2, 1)\,(x_1, x_2) = c\,x$

s.t. $\begin{bmatrix} 1 & 2 \\ 1 & 1 \\ 1 & -1 \\ 1 & -2 \end{bmatrix} \begin{bmatrix} x_1 \\ x_2 \end{bmatrix} \leq \begin{bmatrix} 10 \\ 6 \\ 2 \\ 1 \end{bmatrix}$

or $AX \leq b$

The dual of the given problem is

Mini. $Z_D = b'\, w$

$= [10, 6, 2, 1]\, [w_1, w_2, w_3, w_4]$

$Z_D = 10w_1 + 6w_2 + 2w_3 + w_4$

s.t. $A'\, w \geq c'$

or $\begin{bmatrix} 1 & 1 & 1 & 1 \\ 2 & 1 & -1 & -2 \end{bmatrix} \begin{bmatrix} w_1 \\ w_2 \\ w_3 \\ w_4 \end{bmatrix} \geq \begin{bmatrix} 2 \\ 1 \end{bmatrix}$

$w_1 + w_2 + w_3 + w_4 \geq 2$

$2w_1 + w_2 - w_3 - 2w_4 \geq 1$

$w_1, w_2, w_3, w_4 \geq 0$

Example 5:

Write the dual of the following L.P.P.

Max. $Z = 2x_1 + 3x_2 + x_3$

subject to $4x_1 + 3x_2 + x_3 = 6$

$x_1 + 2x_2 + 5x_3 = 4,$

$x_1, x_2, x_3 \geq 0.$

Solution:

First we shall convert the given L.P.P. into standard primal form.

Since the given problem is of maximization, so all the constraints should have the sign $\leq$.

The standard primal form of the given L.P.P. is

Max. $Z = 2x_1 + 3x_2 + x_3$

subject to $4x_1 + 3x_2 + x_3 \leq 6$

$-4x_1 - 3x_2 - x_3 \leq -6$

$x_1 + 2x_2 + 5x_3 \leq 4$

$-x_1 - 2x_2 - 5x_3 \leq -4,$

$x_1, x_2, x_3 \geq 0.$

The matrix form of the above problem is

Max. $Z = (2, 3, 1)\,(x_1, x_2, x_3)$

subject to

$$\begin{bmatrix} 4 & 3 & 1 \\ -4 & -3 & -1 \\ 1 & 2 & 5 \\ -1 & -2 & -5 \end{bmatrix} \begin{bmatrix} x_1 \\ x_2 \\ x_3 \end{bmatrix} \leq \begin{bmatrix} 6 \\ -6 \\ 4 \\ -4 \end{bmatrix},$$

Now the dual of the given primal is

Min. $Z_D = (6, -6, 4, -4)\,(y_1', y_1'', y_2', y_2'')$

$= 6\,(y_1' - y_1'') + 4\,(y_2' - y_2'')$

subject to

$$\begin{bmatrix} 4 & -4 & 1 & -1 \\ 3 & -3 & 2 & -2 \\ 1 & -1 & 5 & -5 \end{bmatrix} \begin{bmatrix} y_1' \\ y_1'' \\ y_2' \\ y_2'' \end{bmatrix} \geq \begin{bmatrix} 2 \\ 3 \\ 1 \end{bmatrix}.$$

or $4\,(y_1' - y_1'') + y_2' - y_2'' \geq 2$

$3\,(y_1' - y_1'') + 2\,(y_2' - y_2'') \geq 3$

$(y_1' - y_1'') + 5\,(y_2' - y_2'') \geq 1,$

$y_1', y_1'', y_2', y_2'' \geq 0.$

Hence the required dual is

Max. $Z_D = -5y_1 - 3y_2 + 4y_3$

subject to $-y_1 - y_2 \leq 1$

$3y_1 + 2y_2 + 2y_3 \leq 1$

$-4y_1 - y_3 = 1,$

$y_2, y_3 \geq 0$ y_1 is unrestricted.

Substituting $y_1 = y_1' - y_1''$, $y_2 = y_2 - y_2''$, the required dual is

Min $Z_D = 6y_1 + 4y_2$

subject to $4y_1 + y_2 \geq 2$

$3y_1 + 2y_2 \geq 3$

$y_1 + 5y_2 \geq 1.$

y_1, y_2 are unrestricted in sign.

Example 6:

Write the dual of the following problem

$$\text{Max. } Z = 3x_1 + 5x_2 + 7x_3$$

subject to $x_1 + x_2 + 3x_3 \le 10$

$$4x_1 - x_2 + 2x_3 \ge 15,$$

$$x_1, x_2 \ge 0, \; x_3 \text{ is unrestricted.}$$

Solution:

First we shall write the given L.P.P. in the standard primal form.

Since the given problem is of maximization therefore all the constraints should have the sign $\le$.

Substituting $x_3 = x_3' - x_3''$, the standard primal form of the problem is

$$\text{Max. } Z = 3x_1 + 5x_2 + 7(x_3' - x_3'')$$

subject to

$$x_1 + x_2 + 3x_3' - 3x_3'' \le 10$$

$$-4x_1 + x_2 - 2x_3 + 2x_3'' \le -15$$

$$x_1, x_2, x_3', x_3'', \ge 0.$$

The matrix form of the above problem is

$$\text{Max. } Z = (3, 5, 7, -7)(x_1, x_2, x_3', x_3'')$$

subject to

$$\begin{bmatrix} 1 & 1 & 3 & -3 \\ -4 & 1 & -2 & 2 \end{bmatrix} \begin{bmatrix} x_1 \\ x_2 \\ x_3' \\ x_3'' \end{bmatrix} \le \begin{bmatrix} 10 \\ -15 \end{bmatrix}.$$

Now the dual of the given problem is

$$\text{Min. } Z_D = (10, -15)(y_1, y_2)$$

subject to

$$\begin{bmatrix} 1 & -4 \\ 1 & 1 \\ 3 & -2 \\ -3 & 2 \end{bmatrix} \begin{bmatrix} y_1 \\ y_2 \end{bmatrix} \ge \begin{bmatrix} 3 \\ 5 \\ 7 \\ -7 \end{bmatrix}$$

or $\quad \text{Min. } Z_D = 10y_1 - 15y_2$

subject to

$y_1 - 4y_2 \geq 3$

$y_1 + y_2 \geq 5$

$3y_1 - 2y_2 \geq 7$

$-3y_1 + 2y_2 \geq -7$ or $3y_1 - 2y_2 \leq 7_1$

$y_1, y_2 \geq 0.$

Thus the required dual is

Min. $Z_D = 10y_1 - 15y_2$

$y_1 - 4y_2 \geq 3$

$y_1 + y_2 \geq 5$

$3y_1 - 2y_2 = 7$

$y_1, y_2 \geq 0.$

Example 7:

Consider the symmetric simplex problem

Max. $Z_x = 5x_1 + 9x_2$

subject to $x_1 \leq 6$

$x_1 + x_2 \leq 13$

$x_2 \leq 8,$

$x_1, x_2 \geq 0.$

Solution:

The corresponding dual problem is

Min. $Z_w = 6w_1 + 13w_2 + 8w_3$

subject to $w_1 + w_2 \geq 5$

$w_2 + w_3 \geq 9,$

$w_1, w_2, w_3 \geq 0.$

Note: The following table a simple and convenient method to remember the primal dual relationship:

$(x_1, x_2, \ldots, x_n)$ Min.

$$\begin{bmatrix} w_1 \\ w_2 \\ \vdots \\ w_m \end{bmatrix} \begin{bmatrix} a_{11} & a_{12} & \cdots & a_{1n} \\ a_{21} & a_{22} & \cdots & a_{2n} \\ \cdots & \cdots & \cdots & \cdots \\ a_{m1} & a_{m2} & \cdots & a_{mn} \end{bmatrix} \leq \begin{bmatrix} b_1 \\ b_2 \\ \vdots \\ b_m \end{bmatrix}$$

Max. $(c_1, c_2, \ldots, c_n)$

Variables then it is always possible to obtain the dual of this system. In fact every equation is equal to two inequalities one with less than and second with greater than inequality and every unrestricted variable is equal to difference of two positive variable. A constraint with ≤ sign can be converted into a constraint with ≥ sign by multiply by −1.

Example 8:

Write the dual simplex of the problem.

$$\text{Maxi.} \quad Z = 13x_1 + x_2$$

$$\text{s.t.} \quad 12x_1 + 3x_2 \leq 20$$

$$4x_1 + x_2 \leq 35.$$

Solution:

In matrix form the given problem can be written as

$$\text{Max.} \quad Z = (13, 1)(x_1, x_2)$$

$$= cx$$

$$\text{s.t.} \quad \begin{bmatrix} 12 & 3 \\ 4 & 1 \end{bmatrix} \begin{bmatrix} x_1 \\ x_2 \end{bmatrix} \leq \begin{bmatrix} 20 \\ 35 \end{bmatrix}$$

$$\text{or} \quad ax \leq b.$$

The dual of the given problem is

$$\text{Mini.} \quad Z_D = b'\,w$$

$$= (20, 35)(w_1, w_2)$$

$$\text{Mini.} \quad Z_D = 20w_1 + 35w_2$$

$$\text{s.t.} \quad A'\,w \geq c'$$

$$\begin{bmatrix} 12 & 4 \\ 3 & 1 \end{bmatrix} \begin{bmatrix} w_1 \\ w_2 \end{bmatrix} \geq \begin{bmatrix} 13 \\ 1 \end{bmatrix}$$

$$12w_1 + 4w_2 \geq 13$$

$$3w_1 + w_2 \geq 1$$

$$w_1, w_2 \geq 0.$$

Example 9:

Find the dual simplex of the L.P. problem

$$\text{Mini} \quad Z = 4x_1 + 6x_2 + 18x_3$$

$$\text{s.t.} \quad x_1 + 3x_3 \geq 3$$

$$x_2 + 2x_3 \geq 5$$
$$x_1, x_2, x_3 \geq 0.$$

Solution:

In matrix from the given problem can be written as

Mini. $Z = 4, 6, 18) (x_1, x_2, x_3)$

$= c\,x$

s.t. $\begin{bmatrix} 1 & 0 & 3 \\ 0 & 1 & 2 \end{bmatrix} \begin{bmatrix} x_1 \\ x_2 \\ x_3 \end{bmatrix} \geq \begin{bmatrix} 3 \\ 5 \end{bmatrix}$

or $AX \geq b$

The dual of the given problem is

Max. $Z_D = b'\,w$

$= (3, 5)\,(w_1, w_2)$

$= 3w_1 + 5w_2$

Max. $Z_D = 3w_1 + 5w_2$

s.t. $A'\,w \leq c'$

$$\begin{bmatrix} 1 & 0 \\ 0 & 1 \\ 3 & 2 \end{bmatrix} \begin{bmatrix} w_1 \\ w_2 \end{bmatrix} \leq \begin{bmatrix} 4 \\ 6 \\ 18 \end{bmatrix}$$

or $1.w_1 + 0w_2 \leq 4$

$0w_1 + 1w_2 \leq 6$

$3w_1 + 2w_2 \leq 18$

$w_1, w_2 \geq 0.$

Example 10:

The dual simplex method of L.P. problem.

Max. $Z = 2x_1 + 2x_2$

s.t. $2x_1 - 3x_2 \leq 3$

$4x_1 + x_2 \leq -4$

$x_1, x_2 \geq 0.$

Solution:

In matrix from the given problem can be written as

Max. $Z = (2, 2)\,(x_1, x_2)$

$= c\,x$

s.t. $\begin{bmatrix} 2 & -3 \\ 4 & 1 \end{bmatrix} \begin{bmatrix} x_1 \\ x_2 \end{bmatrix} \le \begin{bmatrix} 3 \\ -4 \end{bmatrix}$

or $A\,X \le b$

The dual of the given problem is

Mini $Z_D = b'\,w$

$= (3, -4)\,(w_1, w_2)$

$Z_D = 3w_1 - 4w_2$

s.t. $A'\,w \ge c'$

$$\begin{bmatrix} 2 & 4 \\ -3 & 1 \end{bmatrix} \begin{bmatrix} w_1 \\ w_2 \end{bmatrix} \ge \begin{bmatrix} 2 \\ 2 \end{bmatrix}$$

$2w_1 + 4w_2 \ge 2$

$-3w_1 + w_2 \ge 2$

$w_1, w_2 \ge 0$

Example 11:

Find the dual simplex method of the following L.P.P.

Max. $Z = 6x_1 + 5x_2 + 10x_3$

s.t. $4x_1 + 5x_2 + 7x_3 \le 5$

$3x_1 + 0x_2 + 7x_3 \le 10$

$2x_1 + x_2 + 8x_3 \le 20$

$0x_1 + 2x_2 + 9x_3 \ge 5$

$x_1, x_3 \ge 0$

x_2 *an restricted in sign.*

Solution:

First of all we shall write the given problem in standard form as follows.

(i) Since it is a maximization problem, all the constraints must contain the sign $\le$.

Therefore we multiply the fourth constraints by -1 we get

$$-2x_2 - 9x_3 \le -5.$$

(ii) The variable x_2 is unrestricted in sign

$\therefore$ we write $x_2 = x_2 - x_2''$ where $x_2', x_2'' \ge 0$

So first constraint is equivalent to

$$4x_1 + 5(x_2' - x_2'') + 7x_3 \le 5$$

The third constraint can be written as

$$2x_1 + (x_2' - x_2'') + 8x_3 \le 20$$

and fourth constraint is equivalent to

$$2(x_2' - x_2'') + 9x_3 \ge 5$$

or $\quad -2x_2' + 2x_2'' - 9x_3 \le -5$

Thus the given problem is standard primal form is

$$\text{Max. } Z = 6x_1 + 5(x_2' - x_2'') + 10x_3$$

$$= (6, 5, -5, 10)\,[x_1, x_2', x_2'', x_3]$$

$$= c\,x$$

s.t.

$$4x_1 + 5(x_2' - x_2'') + 7x_3 \le 5$$

$$3x_1 + 7x_3 \le 10$$

$$2x_1 + x_2' - x_2'' + 8x_3 \le 20$$

$$-2x_2' + 2x_2'' - 9x_3 \le -5$$

or

$$\begin{bmatrix} 4 & 5 & -5 & 7 \\ 3 & 0 & 0 & 7 \\ 2 & 1 & -1 & 8 \\ 0 & -2 & 2 & -9 \end{bmatrix} \begin{bmatrix} x_1 \\ x_2' \\ x_2'' \\ x_3 \end{bmatrix} \le \begin{bmatrix} 5 \\ 10 \\ 20 \\ -5 \end{bmatrix}$$

$$A\,x \le b$$

and $\quad x_1, x_2', x_2'', x_3 \ge 0$

The dual of the given problem is

$$\text{Mini. } Z_D = b'\,w$$

$$= [5, 10, 20, -5]\,[w_1, w_2, w_3, w_4]$$

$$Z_D = 5w_1 + 10w_2 + 20w_3 - 5w_4$$

s.t. $\quad A'\,w \le c'$

$$\begin{bmatrix} 4 & 3 & 2 & 0 \\ 5 & 0 & 1 & -2 \\ -5 & 0 & -1 & 2 \\ 7 & 7 & 8 & -9 \end{bmatrix} \begin{bmatrix} w_1 \\ w_2 \\ w_3 \\ w_4 \end{bmatrix} \le \begin{bmatrix} 6 \\ 5 \\ -5 \\ 10 \end{bmatrix}$$

$$4w_1 + 3w_2 + 2w_3 + 0w_4 \ge 6$$

$$5w_1 + 0w_2 + 1w_3 - 2w_4 \ge 5$$

$-5w_1 + 0w_2 - 1w_3 + 2w_4 \geq -5$

$7w_1 + 7w_2 + 8w_3 - 9w_4 \geq 10$

$w_1, w_2, w_4 \geq 0$

and w_3 is unrestricted.

Example 12:

Find the dual simplex method of following L.P.P.

Mini. $Z = x_1 + x_2 + x_3$

s.t. $x_1 - 3x_2 + 4x_3 = 5$

$x_1 - 2x_2 \leq 3$

$2x_2 - x_3 \geq 4$

$x_1, x_2 \geq 0$, x_3 *is unrestricted in sign.*

Solution:

First of all we shall write the given problem in standard form as follows:

(i) It is a minimization problem, all the constraints must contain the sign $\geq$.

Therefore, we multiply the second constraints by –1 we get

$$-x_1 + 2x_2 \geq -3.$$

(ii) The variable x_3 is unrestricted in sign

$\therefore$ we write $x_3 = x_3' - x_3''$ where $x_3', x_3'' \geq 0$

So first constraint is equivalent to

$x_1 - 3x_2 + 4(x_3' - x_3'') \geq 5$ – (i)

$x_1 - 3x_2 + 4(x_3' - x_3'') \leq 5$ – (ii)

Multiply above (ii) by –1

$$-x_1 + 3x_2 - 4(x_3' - x_3'') \geq -5$$

Thus the given problem is standard primal form is

Mini. $Z = x_1 + x_2 + x_3' - x_3''$

$= (1, 1, 1, -1)\,[x_1, x_2, x_3', x_3'']$

$= c\,x$

s.t. $x_1 - 3x_2 + 4x_3' - 4x_3'' \geq 5$

$-x_1 + 3x_2 - 4x_3' + 4x_3'' \geq -5$

$-x_1 + 2x_2 + 0x_3' - 0x_3'' \geq -3$

$0x_1 + 2x_2 - x_3' + x_3'' \geq 4$

or
$$\begin{bmatrix} 1 & -3 & 4 & -4 \\ -1 & 3 & -4 & 4 \\ -1 & 2 & 0 & 0 \\ 0 & 2 & -1 & 1 \end{bmatrix} \begin{bmatrix} x_1 \\ x_2 \\ x_3 \\ x_3 \end{bmatrix} \geq \begin{bmatrix} 5 \\ -5 \\ -3 \\ 4 \end{bmatrix}$$

or $\quad A\,x \geq b$

and $\quad x_1, x_2, x_3', x_3'' \geq 0$

$\therefore$ The dual of the given problem is

Max. $\quad Z_D = b'\,w$

$\quad = (5, -5, -3, 4)\,(w_1', w_1'', w_2, w_3)$

$\quad = 5w_1' - 5w_1'' - 3w_2 + 4w_3$

s.t. $\quad A'w \geq c'$

$$\begin{bmatrix} 1 & -1 & -1 & 0 \\ -3 & 3 & 2 & 2 \\ 4 & -4 & 0 & -1 \\ -4 & 4 & 0 & 1 \end{bmatrix} \begin{bmatrix} w_1' \\ w_1'' \\ w_2 \\ w_3 \end{bmatrix} \leq \begin{bmatrix} 1 \\ 1 \\ 1 \\ -1 \end{bmatrix}$$

or $w_1' - w_1'' - w_2 + 0.w_3 \leq 1$

$-3w_1' + 3w_1'' + 2w_2 + 2w_3 \leq 1$

$4w_1' - 4w_1'' + 0.w_2 - w_3 \leq 1$

$-4w_1' + 4w_1'' + 0.w_2 + w_3 \leq -1$

and $\quad w_1', w_1'', w_2, w_3 \geq 0$

writing $w_1' - w_1'' = w_1$, the dual problem is also written as

Max. $\quad Z_D = 5w_1 - 3w_2 - 4x_3$

s.t. $\quad w_1 - w_2 \leq 1$

$-3w_1 + 2w_2 + 2w_3 \leq 1$

$4w_1 - w_3 = 1$

$w_2, w_3 \leq 0$, w_1 unrestricted in sign.

Example 13:

Apply the principle of Dual problem to solve the following linear programming problem.

Min. $\quad Z_D = 2x_1 + 2x_2$

s.t. $\quad 2x_1 + 4x_2 \geq 1$

$$2x_1 + x_2 \geq 1$$
$$x_1 + 2x_2 \geq 1$$
$$x_1, x_2 \geq 0.$$

Solution:

The dual of this problem is given by

Max. $Z_D = w_1 + w_2 + w_3$

s.t. $2w_1 + 2w_2 + w_3 \leq 2$

$4w_1 + w_2 + 2w_3 \leq 3$

Introducing the slack variable w_{1s} and w_{2s}. The dual problem can be written as

Max. $Z_D = w_1 + w_2 + w_3$

s.t. $2w_1 + 2w_2 + w_2 + w_{1s} = 2$

$4w_1 + w_2 + 2w_3 + w_{2s} = 3$

$w_1, w_2, w_{1s}, w_{2s} \geq 0$

Taking $w_1 = 0 = w_2 = w_3$, we have $w_{1s} = 2$ $w_{2s} = 2$. Which is the starting basic feasible solution of the dual.

First Simplex Table

		c_J	1	1	1	0	0	w_B
D	c_B	w_B	Y_1	Y_2	Y_3	Y_{1s}	Y_{2s}	Y_1
x_{1s}	0	2	2	2	1	1	0	1
x_{2s}	0	2	4	1	2	0	1	1/2 mini
		$Z_J - c_J$	–1	–1	–1	0	0	

Proceeding as usual the second simplex table.

Second Table

		c_J	1	1	1	0	0	w_B
B	c_B	w_B	Y_1	Y_2	Y_3	Y_{1s}	Y_{2s}	y
x_{1s}	0	1	0	3/2	0	1	–1/2	x
x_1	1	1/2	1	1/4	1/2	0	1/4	1 mini
			0	–3/4	–1/2	0	1/4	

Proceeding as usual third simplex table.

Third Table

		c_J	1	1	1	0	0	w_B
B	c_B	w_B	Y_1	y_2	Y_3	Y_{1s}	Y_{2s}	y_2
x_{1s}	0	1	0	3/2	0	1	–1/2	2/3 Mini
x_3	1	1	2	1/2	1	0	1/2	2
	$Z_J - c_J$		1	–1/2	0	0	1/2	
				↑		↓		

Proceeding as usual fourth simplex table.

Fourth Table

		c_J	1	1	1	0	0
B	c_B	w_B	Y_1	Y_2	Y_3	Y_{1s}	Y_{2s}
x_2	1	2/3	0	1	0	2/3	–1/3
x_3	1	2/3	2	0	1	–1/3	2/3
	$Z_J - c_J$		1	0	0	1/3	1/3

Since all $Z_J - c_J \geq 0$. So solution is optimal.

Solution of primal problem

$x_1 = 1/3$

$x_2 = 1/3$

$Z = 4/3$

For primal constraints, read across the table while for dual constraints read down the columns.

Example 14:

Find the dual of problem

$$\begin{aligned} \textit{Max.}\quad & Z = -x_1 + 2x_2 - x_3 \\ \textit{s.t.}\quad & 3x_1 + x_2 - x_3 \leq 10 \\ & -x_1 + 4x_2 + x_3 \geq 6 \\ & x_2 + x_3 \leq 4 \\ & x_1, x_2, x_3 \geq 0 \end{aligned}$$

Solve the primal problem by simplex method and deduce the solution of the problem from the optimal tableaus of the primal.

Solution:

In standard form the problem can be written as

$$\text{Max. } Z_P = -x_1 + 2x_2 - x_3$$

$$3x_1 + x_2 - x_3 \le 10$$

$$x_2 - 4x_2 - x_3 \le 6$$

$$x_2 + x_3 \le 4$$

$$x_1, x_2, x_3, x_4 \ge 0$$

The dual of this problem can be written as

$$\text{Mini. } Z_D = 10w_1 + 6w_2 + 4w_3$$

$$3w_1 \ge -1$$

$$w_1 + w_2 + w_3 \ge 2$$

$$-w_1 - w_2 + w_3 \ge -1$$

Introducing the slack variable in first and third constraint and surplus and artificial is second constraint of the primal problem. The problem can be written as

$$3x_1 + x_2 - x_3 + x_{1s} = 10$$

$$-x_1 + 4x_2 + x_3 - x_{2s} + x_{2a} = 6$$

$$x_2 + x_3 + x_{3s} = 4$$

$$x_1, x_2, x_3, \ldots x_{3s} \ge 0$$

Taking $x_1 = 0 = x_2 = x_3 = x_{2s}$,

we have $x_{1s} = 10$,

$x_{2a} = 6,\ x_{3s} = 4$,

which is the starting basic feasible solution of the primal.

First Table

B	c_B	c_J / x_B	-1 / Y_1	2 / Y_2	-1 / Y_3	0 / Y_{1s}	0 / Y_{2s}	0 / Y_{3s}	$-M$ / Y_{2a}	x_B / Y_2
x_{1s}	0	10	3	1	−1	1	0	0	0	10
x_{2a}	−M	6	−1	4	1	0	−1	0	1	3/2 mini→
x_{3s}	0	4	0	1	1	0	0	1	0	4
		Z_J-c_J	M+1	−4+2 ↑	−M+1	0	M	0	0 ↓	

Since $\Delta_2 = -4M - 2$ is mini so y_2 is incoming vector and by mini ratio rule out going vector is y_{2a}.

The key element is $y_{22} = 4$

Proceeding as before the second simplex table

Second Table

B	c_B	c_J / x_B	−1 / Y_1	2 / Y_2	1 / Y_3	0 / Y_{1s}	0 / Y_{2s}	0 / Y_{3s}	−M / Y_{2a}	$\frac{x_B}{Y_{2s}}$
x_{1s}	0	17/2	13/4	0	−5/4	1	1/4	−1/4	0	34
x_2	2	3/2	−1/4	1	1/4	0	−1/4	1/4	0	−ve
x_{3s}	0	5/2	1/4	0	3/4	0	1/4	−1/4	1	10 mini
		$Z_J - c_J$	1/2	0	11/8	0	−1/2	$\frac{1}{2}$ + M		0

Proceeding as before the third simplex table

Third Table

B	c_B	c_J / x_B	−1 / Y_1	2 / Y_2	−1 / Y_3	0 / Y_{1s}	0 / Y_{2s}	0 / Y_{3s}	−M / Y_{2a}
x_{1s}	0	29/4	3	0	−2	1	0	−1	0
x_2	2	4	0	1	1	0	0	1	0
x_{2s}	0	10	1	0	3	0	1	4	−1
	$Z_J - c_J$	1	0	3	0	0	2	M	

Since all $Z_J - c_J \geq 0$, so this solution is optimal.

Optimal solution is

$$x_1 = 0 = x_3, \; x_2 = 4$$

Max. Z = 8

solution of the dual

$$w_1 = 0 = w_2, \; w_3 = 2$$

Mini. Z = 8.

Example 15:

Consider the problem,

Max. $Z = 8x_1 + 6x_2$

s.t. $x_1 - x_2 \le 3/5$

$x_1 - x_2 \ge 2$

$x_1, x_2 \ge 0$

Show that the primal and dual problem have no feasible solution.

Solution:

In standard form the problem can be written as

Max. $Z = 8x_1 + 6x_2$

s.t. $x_1 - x_2 \le 3/5$

$-x_1 + x_2 \le -2$

$x_1, x_2 \ge 0$

The dual of the this problem can be written as

Mini. $Z_D = 3/5\ w_1 - w_2$

$w_1 - w_2 \ge 8$

$-w_1 + w_2 \ge 6$

$w_1, w_2 \ge 0$

Introducing the slack variable in fist constraint and surplus and artificial variable in second constraint of the primal.

The problem can be written as

Max. $Z = 8x_1 + 6x_2$

s.t. $x_1 - x_2 + x_{1s} = 3/5$

$x_1 - x_2 - x_{1s} + x_{2a} = 2$

$x_1, \ldots, x_{2a} \ge 0$

Taking $x_1 = x_2 = x_{2s}$ we have $x_{1s} = 3/5\ x_{2a} = 2$. Which is starting basic feasible solution of the primal.

First Table

		c_J	8	6	0	0	–M
B	c_B	x_B	Y_1	Y_2	Y_{1s}	Y_{2s}	Y_{2a}
x_{1s}	0	3/5	1	–1	1	0	0
x_{2a}	–M	2	1	–1	0	–1	1
		$Z_J–c_J$	–M–8	+M–6	0	M	0

Proceeding as before the second simplex table.

Second Table

B	c_B	c_J / x_B	8 / Y_1	6 / Y_2	0 / Y_{1s}	0 / Y_{2s}	–M / Y_{2a}	$\frac{x_B}{Y_2}$
c_1	8	3/5	1	–1	1	0	0	–ve
x_{2a}	–M	7/5	0	0	–1	–1	1	∞
	$Z_J - c_J$	0	–2 ↑	M+8	M	0		

Every value of $\frac{x_B}{y_2}$ are –ve and ∞, so the primal has no feasible solution and dual has also no feasible solution.

Example 16:

Write the dual of this problem

$$\text{Max. } Z = 2x_1 + 3x_2$$

$$\text{s.t.} \quad 2x_1 + 2x_2 \leq 10$$

$$x_1 + 2x_2 \leq 6$$

$$2x_1 + x_2 \leq 6$$

$$x_1, x_2 \geq 0.$$

Solve the primal by the simplex method.

Solution:

The due of this problem can be written as

$$\text{Mini. } Z_D = 10w_1 + 6w_2 + 6w_3$$

$$2w_1 + w_2 + 2w_3 \geq 2$$

$$2w_1 + 2w_2 + w_3 \geq 3$$

$$w_1, \ldots, w_3 \geq 0$$

Introducing the slack variable x_{1s}, x_{2s}, x_{3s} in the primal problem. The problem can be written as

$$\text{Max. } Z = 2x_1 + 3x_2$$

$$\text{s.t.} \quad 2x_1 + 2x_2 + x_{1s} = 10$$

$$x_1 + 2x_2 + x_{2s} = 6$$

$$2x_1 + x_2 + x_{3s} = 6$$

$$x_1, \ldots, x_{3s} \geq 0$$

Taking $x_1 = 0 = x_2$ have $x_{1s} = 10$, $x_{2s} = 6$, $x_{3s} = 6$

Which is the starting basic feasible solution of the primal.

First Table

		c_J	2	2	0	0	0	$\frac{x_B}{Y_1}$
B	c_B	x_B	Y_1	Y_2	Y_{1s}	Y_{2s}	Y_{3s}	
x_{1s}	0	10	2	2	1	0	0	5
x_{2s}	0	6	1	2	0	1	0	6
x_{3s}	0	6	2	1	0	0	1	2 mini
	$Z_J - c_J$		–2	–2	0	0	0	
			↑				↓	

Proceeding as before the second simplex table.

Second Table

		c_J	2	2	0	0	0	$\frac{x_B}{Y_2}$
B	c_B	x_B	Y_1	Y_2	Y_{1s}	Y_{2s}	Y_{3s}	
x_{1s}	0	4	0	1	1	0	–1	4
x_{2s}	0	3	0	3/2	0	1	–1/2	6
x_1	2	3	1	1/2	0	0	1/2	6
			0	–1	0	0	1	
				↑		↓		

Proceeding as before the third simplex table.

Third Table

		c_J					
B	c_B	x_B	Y_1	Y_2	Y_{1s}	Y_{2s}	Y_{3s}
x_{1s}	0	2	0	2	1	2/3	–2/3
x_2	3	2	0	1	0	2/3	–1/3
x_1	2	2	1	0	0	–1/3	2/3
	$Z_J–c_J$		0	1	0	4/3	1/3

Since all $Z_J - c_J \geq 0$, so this solution is optimal. The solution of the primal problem is

$x_1 = 2 = x_2$

Max $Z_P = 10$

The solution of dual $w_1 = 0$

$w_2 = 4/3$

$w_3 = 1/3$

Mini. $Z_D = 10$

Example 17:

Use Dual problem to solve the L.P.P.

Max. $Z = 5x_1 - 2x_2 + 3x_3$

s.t. $2x_1 + 2x_2 - x_3 \geq 2$

$3x_1 - 4x_2 \leq 3$

$x_2 + 2x_3 \geq 5$

$x_1, x_2, x_3 \geq 0.$

Solution:

In the standard form the problem is written as

Max. $Z = 5x_1 - 2x_2 + 3x_3$

s.t. $2x_1 - 2x_2 + x_3 \leq - 2$

$3x_1 - 4x_2 \leq 3$

$x_2 + 3x_3 \leq 5$

$x_1, x_2, x_3 \leq 0$

The dual of this problem is given by

Min $Z_D = - 2w_1 + 3w_2 + 5w_3$

s.t. $- 2w_1 + 3w_2 \geq 5$

$- 2w_1 - 4w_2 + w_3 \geq - 2$

$w_1 + 3w_3 \geq 3$

$w_1, w_2, w_3 \geq 0$

Multiply the second constraint by $- 1$

$2w_2 + 4w_2 - w_3 \leq 2$

Introducing the surplus and artificial variable in first and third constraint and slack variable in second constraint. The dual problem reduce to the following form.

Max. $Z_D = 2w_1 - 3w_2 - 5w_3$

s.t. $-2w_1 + 3w_2 - w_{1s} + w_{1a} = 5$

$2w_1 + w_2 - w_3 + w_{2s} = 2$

$w_1 + 3w_3 - w_{3s} + w_{3a} = 3$

$w_1, w_2, \ldots, w_{3a} \geq 0$

Taking $w_1 = 0 = w_2 = w_3 = w_{1s} = w_{3s}$, we have $w_{1a} = 5$, $w_{2s} = 2$, $w_{3a} = 3$

Which is the starting basic feasible solution of the dual.

First Table

		c_J	2	–3	–5	0	0	0	–M	–M	w_B
B	c_B	w_B	Y_1	Y_2	Y_3	Y_{1s}	Y_{2s}	Y_{3s}	Y_{1a}	Y_{3a}	Y_2
x_{1a}	–M	5	–2	3	0	–1	0	0	1	0	5/3
x_{2s}	0	2	2	4	–1	0	1	0	0	0	2/4mini→
x_{3a}	–M	3	1	0	3	0	0	–1	0	1	∞
		Z_J-c_J	M–2	–3M+3 ↑	–3M+5	M	0 ↓	M	0	0	

Proceeding as before second simplex table

Second Table

		c_J	2	–3	–5	0	0	0	–M	–M	w_B
B	c_B	w_B	Y_1	Y_2	Y_3	Y_{1s}	Y_{2s}	Y_{3s}	Y_{1a}	Y_{2a}	Y_3
x_{1a}	–M	7/2	–7/2	0	3/4	–1	–3/4	0	1	0	14/3
x_2	–3	1/2	1/	1	–1/4	0	1/4	0	0	0	–ve
x_{3s}	–M	3	1	0	3	0	0	–1	0	1	1 mini
		Z_J-c_J	$\frac{5M-7}{2}$	0	$\frac{23-15M}{2_1}$ ↑	M	$\frac{3M-3}{4}$	M	0	0 ↓	

Proceeding as usual third simplex table

Third Table

		c_J	2	–3	–5	0	0	0	–M	–M	w_B
B	c_B	w_B	Y_1	Y_2	Y_3	Y_{1s}	Y_{2s}	Y_{3s}	Y_{1a}	Y_{3a}	
x_{1a}	–M	11/4	–15/4	0	0	–1	–3/4	(1/4)	1	–1/4	11 mini
x_2	–3	3/4	7/12	1	0	0	1/4	–1/12	0	1/12	–ve
x_3	–5	1	1/3	0	1	0	0	–1/3	0	1/3	–ve

$\frac{45M-95}{12}$ 0 0 −M $\frac{3M-3}{4}$ 0 $\frac{5M-6}{4}$
↑ ↓

Proceeding as usual fourth simplex table

Fourth Table

		c_J	2	−3	−5	0	0	0
B	c_B	w_B	y_1	Y_2	Y_3	Y_{1s}	Y_{2s}	Y_{3s}
x_{3s}	0	11	−15	0	0	−4	−3	1
x_2	−3	5/3	−2/3	1	0	−1/3	0	0
x_3	−5	14/3	−14/3	0	1	−4/3	−1	0
		$Z_J - c_J$0	0	0	23/3	5	0	

Since all $Z_J - c_J \geq 0$, so the solution is optimal.

∴ Solution of the primal problem is

$x_1 = 23/3 \; x_3 = 0$

$x_2 = 5$

Max. $Z_P = \frac{85}{3}$.

Example 18:

Using the dual, solve the L.P.P.

Mini. $Z_P = 3x_1 + x_2$

s.t. $2x_1 + 3x_2 \geq 2$

$x_1 + x_2 \geq 1$

$x_1, x_2, \geq 0.$

Solution:

The dual of the given L.P.P. is

Max. $Z_D = 2w_1 + w_2$

s.t. $2w_1 + w_2 \leq 3$

$3w_1 + w_2 \leq 1$

$w_1, w_2, \geq 0$

Introducing the slack variable w_{1s} and w_{2s} respectively. The dual problem can be written as

Max. $Z_D = 2w_1 + w_2$

s.t. $2w_1 + w_2 + w_{1s} = 3$

$3_{w1} = w_2 + w_{2s} = 1$

Taking $w_1 = 0 = w_2$, we have $w_{1s} = 3$ and $w_{2s} = 1$. Which is the starting basic feasible solution of the dual.

First Simplex Table

		c_J	2	1	0	0	w_B
B	c_B	w_B	Y_1	Y_2	Y_{1s}	Y_{2a}	Y_{i1}
x_{1s}	0	3	2	1	1	0	3/2
x_{2s}	0	1	3	1	0	1	1/3→mini
	Z_J	$-c_J$	–2	–1	0	0	
	↑					↓	
	incoming vector					outgowing vector	

$\Delta_1 = Z_1 - c_1 = c_B\, Y_1 - c_1 = [(0 \times 2 = 0 \times 3) - 2] = -2$
$\Delta_2 = Z_2 - c_2 = c_B\, y_2 - c_2 = [0 \times 1 + 0 \times 1] - 1 = -1$
$\Delta_3 = Z_3 - c_3 = c_B\, y_3 - c_3 = 0$
$\Delta_4 = 0$

To find incoming vector

Since $\Delta_1 = -1$ is mini. of $\Delta_1, \Delta_2, \Delta_3, \Delta_4$. So y_1 is incoming vector 0.

To find outgoing vector

Since y_1 is incoming vector, so we find ratio

$$= \text{Mini}\left\{\frac{w_B}{y_{i1}}\right\}$$

$$= \text{Mini.}\left\{\frac{w_B}{y_{11}}, \frac{w_B}{y_{21}}\right\}$$

$$= \text{Mini} \quad \left\{\frac{3}{2}, \frac{1}{3}\right\}$$

$\frac{1}{3}$ is mini so x_{2s} is out going vector.

The key element is $y_{21} = 3$.

Firstly, we divide the second marked row of the first table by 3 to get the second row of the second table.

Now, Multiply the new second row by 2 (number in first row and marked column of the first table).

This will give

$\frac{1}{3}\times 2 \quad 1\times 2 \quad \frac{1}{3}\times 2 \quad 0\times 2 \quad \frac{1}{3}\times 2$

$\frac{2}{3} \quad 2 \quad \frac{2}{3} \quad 0 \quad \frac{2}{3}.$

Now subtracting these from the corresponding element of the first row of the first table. This will give

$3-\frac{2}{3} \quad 2-2 \quad 1-\frac{2}{3} \quad 1-0 \quad 0-\frac{2}{3}$

$\frac{7}{3} \quad 0 \quad \frac{1}{3} \quad 1 \quad -\frac{2}{3}.$

This is the first row of the second table.

Second Table

		c_J	2	1	0	0	w_B
B	c_B	w_B	Y_1	Y_2	Y_{1s}	Y_{2s}	Y_{12}
x_{1s}	0	7/3	0	1/3	1	–2/3	7
x_1	2	1/3	1	1/3	0	1/3	1 mini→
			0	–1/3	0	2/3	
			↓		↑		
			outgoing vector		incoming vector		

$$\Delta_1 = Z_1 - c_1 = 0$$
$$\Delta_2 = Z_2 - c_2 = -1/3$$
$$\Delta_3 = Z_3 - c_3 = 0$$
$$\Delta_4 = Z_4 - c_4 = 2/3$$

To find incoming vector

Since $\Delta_2 = -\frac{1}{3}$ is mini of all. so y_2 is incoming vector

To find out vector.

Since y_2 is incoming vector. So we will find ratio Mini. $\left\{\frac{w_B}{y_{i2}}\right\}$

$$= \text{Mini}\left\{\frac{w_B}{y_{12}}, \frac{w}{y_{22}}\right\}$$

$$= \text{Mini}\left\{\frac{7\times 3}{3\times 1}, \frac{1\times 3}{3\times 1}\right\}$$

$$= \text{Mini}\ \{7, 1\}$$

1 is mini so x_1 is out going vector.

The key element is $y_{22} = 1/3$.

First we divide the second marked row of the second table by 1/3 to get the new second row of the third table.

Now multiply the new second row by 1/3 (number in the first row and marked column of the second table.

This will give

$$1\times\frac{1}{3} \qquad 3\times\frac{1}{3} \qquad 1\times\frac{1}{3} \qquad 0\times\frac{1}{3} \qquad 1\times\frac{1}{3}$$

$$\frac{1}{3} \qquad 1 \qquad \frac{1}{3} \qquad 0 \qquad \frac{1}{3}$$

Now subtracting these from the corresponding element of the first row of the second table. This will give

$$\frac{7}{3}-\frac{1}{3} \qquad 0-1 \qquad \frac{1}{3}-\frac{1}{3} \qquad 1-0 \qquad \frac{-2}{3}\frac{-1}{3}$$

$$2 \qquad -1 \qquad 0 \qquad 1 \qquad -1.$$

This is the first row of the third table.

Third Table

		c_J	2	1	0	0
B	c_B	w_B	Y_1	Y_2	Y_{1s}	Y_{2s}
x_{1s}	0	2	–1	0	1	–1
x_2	1	1	3	1	0	1
		$Z_J - c_J$	1	0	0	1

Since all $Z_J - c_J \geq 0$, so this solution is optimal.

Optimal solution of the dual problem is

$$w_1 = 0$$

$$w_2 = 1$$

$$\text{Max } Z_D = 1$$

solution of original problem

Form the find table solution of the primal is given by

$$x_1 = 0$$

$$x_2 = 1$$

$$\text{Mini. } Z_P = \text{Max } Z_D = 1$$

Example 19:

Using the dual, solve the following L.P.P.

$$\text{Max} \quad Z_P = 3x_1 - 2x_2$$

$$\text{s.t.} \quad x_1 \le 4$$

$$x_2 \le 6$$

$$x_1 + x_2 \le 5$$

$$-x_2 \le -1$$

$$x_1, x_2 \ge 0.$$

Solution:

The dual of the given problem is given by

$$\text{Mini} \quad Z_D = 4w_1 + 6w_2 + 5w_3 - w_4$$

$$\text{s.t.} \quad w_1 + w_3 \ge 3$$

$$w_2 + w_3 - w_4 \ge - 2$$

$$\text{and} \quad w_1, w_2, w_3, w_4 \ge 0.$$

Now firstly we cell convert the dual problem of minimization to maximization problem by taking the objective function as

$$\text{Max} \quad Z_D' = - Z_D = - 4w_1 - 6w_2 - 5w_3 + w_4$$

$$\text{s.t.} \quad w_1 + w_3 \ge 3$$

$$-w_2 - w_3 + w_4 \le 2$$

Introducing the surplus, slack variable w_{1s}, w_{2s}, respectively, the constraints of the dual problem reduces to the following equities.

$$w_1 + w_3 - w_{1s} = 3$$

$$-w_2 - w_3 + w_4 + w_{2s} = 2$$

$$w_1, w_2, w_3, w_4, w_{1s}, w_{2s} \ge 0.$$

Since identity matrix I_2 is present in the coefficient matrix. So there is no need to introduce artificial variable in the first constraint.

Taking $w_2 = 0 = w_3 = w_4 = w_{1s}$, we have $w_1 = 3$ and $w_{2s} = 2$. Which is the starting basic feasible solution of the dual.

First Table

		c_J	-4	-6	-5	1	0	0	w_B
B	c_B	w_B	Y_1	Y_2	Y_3	Y_4	Y_{1s}	Y_{2s}	Y_{14}
x_1	-4	3	1	0	1	0	-1	0	$\frac{3}{0} = \infty$
x_{2s}	0	2	0	-1	-1	1	0	1	2 mini→
	$Z_J - c_J$		0	6	1	-1	4	0	
						↑		↓	

Calculate $Z_J - c_J$

$\Delta_1 = Z_1 - c_1 = 0, \Delta_2 = 6, \Delta_3 = 1$

$\Delta_4 = -1, \Delta_5 = 4.$

Since $\Delta_4 = -1$ is mini of $\Delta_1\ \Delta_2\ \Delta_3$ so y_4 is incoming vector.

To find the out going vector since y_4 is incoming vector so we will find ratio Mini $\left\{\frac{w_B}{y_{i4}}, y_{i4} >\right\}$

Mini $\{\infty, 2\}$

2 is mini x_{2s} is out going vector.

The key element is $y_{24} = 1$

Proceeding as before the second simplex table is

Second Simplex Table

		c_J	-4	-6	-5	1	0	0
c_B	w_B	Y_1	Y_2	Y_3	Y_4	Y_{1s}	Y_{2s}	
x_1	-4	3	1	0	1	0	-1	0
x_4	1	2	0	-1	-1	1	0	1
	$Z_J - c_J$		0	5	0	0	4	1

Since all $Z_J - c_J \geq 0$, so this solution of the dual problem is optimal

Optimal solution of the dual problem is

$w_1 = 3, w_2 = 0$

$w_3 = 0, w_4 = 2$

Mini $Z_D = \text{Max } Z_D' = 10$

Solution of the original problem

$x_1 = 4$

$x_2 = 1$

Max. $Z_P = 10.$

Example 20:

Find the dual of the following problem and hence solve it.

Mini $Z_P = 6x_1 + 5x_2 + 2x_3$

s.t. $x_1 + 3x_2 = 2x_3 \geq 5$

$4x_1 - 2x_2 + 3x_3 \geq -1$

$2x_1 + 2x_2 + x_3 \geq 2$

$x_1, x_2, x_3 \geq 0.$

Solution:

The dual of this problem is given by

Max. $Z_D = 5w_1 - w_2 + 2w_2$

s.t. $w_1 + 4w_2 + 2w_3 \leq 6$

$3w_1 - 2w_2 + 3w_3 \leq 5$

$2w_1 + 3w_2 + w_3 \leq 2$

$w_1, w_2, w_3 \leq 0.$

Introducing the slack variable w_{1s}, w_{2s}.

Example 21:

Using the dual solve the L.P.P.

Mini $Z_P = 10x_1 + 6x_2 + 2x_3$

s.t. $-x_1 + x_2 + x_3 \geq 1$

$3x_1 + x_2 - x_3 \geq 2$

$x_1, x_2, x_3, x_4 \geq 0.$

Solution:

The dual of this problem is given by

Max $\quad Z_D = w_1 + 2w_2$

s.t. $\quad -w_1 + 3w_2 \leq 10$

$w_1 + w_2 \leq 6$

$w_1 - w_2 \leq 2$

$w_1, w_2 \geq 0.$

Introducing the slack variable w_{1s}, w_{2s} and w_{3s} respectively. The dual problem can be written as

Max $\quad Z_D = w_1 + 2w_2$

s.t. $\quad -w_1 + 3w_2 + w_{1s} = 10$

$w_1 + w_2 + w_{2s} = 6$

$w_1 - w_2 + w_{3s} = 2$

Taking $w_1 = 0 = w_2$, we have $w_{1s} = 10$, $w_{2s} = 6$ and $w_{3s} = 2$. Which is the starting basic feasible solution of the dual

First Simplex Table

		c_J	1	2	0	0	0	w_B
B	c_B	w_B	Y_1	Y_2	Y_{1s}	Y_{2s}	Y_{3s}	Y_{12}
x_{1s}	0	10	–1	3	1	0	0	10/3mini→
x_{2s}	0	6	1	1	0	1	0	6/1
x_{3s}	0	2	1	–1	0	0	1	–ve
			–1	–2	0	0	0	
				↑	↓			

Calculate of $Z_J - c_J$

$\Delta_1 = Z_1 - c_1 = -1,$

$\Delta_2 = -2$

To find incoming vector

Since $\Delta_2 = -2$ is mini of Δ_1 and Δ_2.

So y_2 is incoming vector.

To find out going vector.

Since y_2 is incoming vector so we will find ratio Mini $\left\{\frac{w_B}{y_{i2}}, y_{i2} > 0\right\}$

$$= \text{Mini}\left\{\frac{w_{B1}}{y_{12}}, \frac{w_{B2}}{y_{22}}, \frac{w_{B3}}{y_{32}}\right\}$$

$$= \text{Mini}\left\{\frac{10}{3}, \frac{6}{1}, \frac{2}{-1}\right\}$$

$\frac{10}{3}$ is mini so x_{1s} in out going.

The key element is $y_{12} = 3$

Firstly, we divide the first marked row of the first table by 3 to get new first row of the second table. Calculating second and third row from first row.

Second Table

		c_J	1	2	0	0	0	w_B
B	c_B	w_B	Y_1	Y_2	Y_{1s}	Y_{2s}	Y_{3s}	y_{11}
x_2	2	10/3	–1/3	1	1/3	0	0	–ve
x_{2s}	0	8/3	4/3	0	–1/3	1	0	2Mini→
x_{3s}	0	16/3	2/3	0	1/3	0	1	8
			–5/3	0	2/3	0	0	
			↑			↓		

calculate of $Z_J - c_J$

$\Delta_1 = Z_1 - c_1 = -5/3,$

$\Delta_2 = Z_2 - c_2 = 0$

$\Delta_2 = 2_3 - c_3 = 2/3,$

To find incoming vector

Since $\Delta_1 = -5/3$ is mini of Δ_1, Δ_2 Δ_3. So y_1 is incoming vector

To find out going vector.

Since y_1 is incoming vector so we will find ratio Mini $\left\{\frac{w_B}{y_{11}}, y_{11} > 0\right\}$

$$= \text{Mini}\left\{\frac{w_{B1}}{y_{11}}, \frac{w_{B2}}{y_{21}}, \frac{w_{B3}}{y_{32}}\right\}$$

$$= \text{Mini}\left\{-\text{ve}, \frac{8\times 3}{3\times 4,}\ \frac{16\times 3}{3\times 2}\right\}$$

2 is mini so x_{2s} is out going vector.

The key element is $y_{21} = 4/3$

First we divide the second marked row of the second table by 4/3 to get the new second of the third table.

Calculating first and third row from second row.

Third Simplex Table

		c_J	1	2	0	0	0
B	c_B	w_B	Y_1	Y_2	Y_{1s}	Y_{2s}	Y_{3s}
x_2	2	4	0	1	1/4	1/4	0
x_1	1	2	1	0	–1/4	3/4	0
x_{3s}	0	2_1	0	0	1/2	–1/2	1
			0	0	1/4	5/4	0

All $Z_J - c_J \geq 0$, so this solution is optimal.

Optimal solution of the dual problem is

$$w_1 = 2$$

$$w_2 = 4$$

$$\text{Max } Z_D = 10$$

Solution of the original problem is

$$x_1 = 1/4$$

$$x_2 = 5/4$$

$$x_3 = 0$$

$$\text{Mini } Z_P = 10.$$

Example 22:

Write the dual of following problem and solve it.

$$\text{Max. } Z = 4x_1 + 2x_2$$

subject to $-x_1 - x_2 \leq -3$

$-x_1 + x_2 \leq -2,\ x_1,\ x_2 \geq 0.$

Hence or otherwise write down the solution of the primal.

Solution:

The given problem is in standard primal form.

$\therefore$ the dual to the given primal is

$$\text{Min. } Z_D = -3w_1 - 2w_2$$

subject to $-w_1 - w_2 \geq 4$

$-w_1 + w_2 \geq 2$

$w_1, w_2 \geq 0.$

Changing the objective function to maximization and introducing surplus variables $w_3 \geq 0$, $w_4 \geq 0$ and artificial variables $w_5, w_6 \geq 0$ the above problem reduces to

$$\text{Max. } Z_D = 3w_1 + 2w_2 + 0w_3 + 0w_4 - Mw_5 - Mw_6$$

subject to $-w_1 - w_2 - w_3 + w_5 = 4$

$-w_1 + w_2 - w_4 + w_6 = 2$

The starting B.F.S. is $w_1 = 0 = w_2 = w_3 = w_3$, $w_5 = 4$, $w_6 = 2$.

Now applying the simplex method to obtain the optimal solution, we have

		c_j	3	2	0	0	–M	–M	Min. ratio
B	c_B	w_B	W_1	W_2	W_3	W_4	W_5	W_6	w_B/W_2
W_5	–M	4	–1	–1	–1	0	1	0	–ive
W_6	–M	2	–1	1	0	–1	0	1	2
$Z_D = -6_M$		Δ_j	3 – 2M	2	–M	–M	0	0	
W_5	–M	6	–2	0	–1	–1	1	1	
W_2	2	2	–1	1	0	–1	0	1	
$Z_D = 4 - 6M$		Δ_j	5 – 2M	0	–M	2 – M	0	–2	

In the last simplex table no $\Delta_j > 0$ and a non-zero artificial variable appears in the basis therefore the dual problem does not possess any optimum basic feasible solution. Consequently, the given problem does not possess any finite optimal solution.

And w_{3s}. The dual problem can be written as

$$\text{Max. } Z_D = 5w_1 - w_2 + 2w_2$$

s.t. $\quad w_1 + 4w_2 + 2w_3 + w_{1s} = 6$

$3w_1 - 2w_2 + 3w_3 + w_{2s} = 5$

$2w_1 + 3w_2 + w_3 + w_{3s} = 2$

$w_1, w_2 - w_{3s} \geq 0.$

Taking $w_1 = 0\ w_2 = w_3$, we have $w_{1s} = 6$, $w_{2s} = 5$, $w_{3s} = 2$ which is the starting basic feasible solution of the dual.

First Simplex Table

B	c_B	c_J / w_B	5 / Y_1	–1 / Y_2	2 / Y_3	0 / Y_{1s}	0 / Y_{2s}	0 / Y_{3s}	w_B / Y_{11}
x_{1s}	0	6	1	4	2	1	0	0	6/1
x_{2s}	0	5	3	–2	2	0	1	0	5/3
x_{3s}	0	2	2	3	1	0	0	1	1 mini
			–5	1	–2	0	0	0	
			↑					↓	

Calculate of $Z_J - c_J$

$\Delta_1 = -5,\ \Delta_2 = 1,$

$\Delta_3 = -2$

Since $\Delta_1 = -5$ is mini of $\Delta_1, \Delta_2, \Delta_3$ so y_1 is incoming vector

To find out going vector

Since y_1 is incoming vector so we will find ratio Mini. $\left\{\frac{w_B}{y_{i1}}, y_{i1} > 0\right\}$

Mini. $\left\{\frac{6}{1}, \frac{5}{3}, \frac{2}{2}\right\}$

1 is mini so y_{3s} is out going vector.

The key element is $y_{31} = 2$

Proceeding as before the second simplex table

Second Table

B	c_B	c_J / w_B	5 / Y_1	–1 / y_2	2 / Y_3	0 / Y_{1s}	0 / Y_{2s}	0 / Y_{3s}
x_{1s}	0	5	0	5/2	3/2	1	0	–1/2
x_{2s}	0	2	0	13/2	1/2	0	1	–3/2
x_1	5	1	1	3/2	1/2	0	0	1/2
	$Z_J - C_J$		0	17/2	5/2	0	0	5/2

Since all $Z_J - c_J \geq 0$. Therefore solution is optimal.

i.e., optimal solution of the dual problems

$w_1 = 1,$

$w_2 = 0,$

$w_3 = 0$

Max. $Z_D = 5$

solution of the primal problem is

$x_1 = 0, x_2 = 0$

$x_3 = 5/2$

Mini. $Z_P = 5.$

Example 23:

Formulate the following L.P.P., into, its dual problem and solve it

Max $Z = x_1 - x_2$

s.t., $2x_1 + x_2 \geq 2$

$- x_1 - x_2 \geq 1$

and $x_1, x_2 \leq 0.$

Solution:

The given L.P.P. is maximizations so all sign should be less than

$- 2x_1 - x_2 \leq - 2$

$x_2 + x_2 \leq - 1$

The dual of this problems is given by

Mini $Z_D = - 2w_1 - w_2$

s.t., $- 2w_1 + w_2 \geq 1$

$- w_1 + w_2 \geq - 1$

Now firstly we will convert the dual problem of minimization to maximizations problem by taking the objective function as

Max $Z_D' = - Z_D = 2w_1 + w_2$

s.t., $- 2w_1 + w_2 \geq 1$

$w_1 - w_2 \leq 1$

Introducing the surplus, slack and artificial variable w_{1s}, w_{2s}, w_{1a} respectively. The dual problem can be written as

Max $Z_D' = - Z_D = 2w_1 + w_2$

s.t., $- 2w_1 + w_2 - w_{1s} + w_{1a} = 1$

$w_2 - w_2 + w_{2s} = 1$

Taking $w_1 = 0 = w_2 = w_{1s}$, we have $w_{1a} = 1$ $w_{2s} = 1$, which is the starting B.F.S.

First Table

		c_J	2	1	0	0	–M	w_B
B	c_B	w_B	Y_1	Y_2	Y_{1s}	Y_{2s}	Y_{1a}	Y_{i2}
x_{1a}	–M	1	–2	1	–1	0	1	1
x_{2s}	0	1	1	–1	0	1	0	–ive
	$Z_J–c_J$		$2_M–2$	–M–1	M	0	0	
			↑				↓	

$\Delta_1 = 2M - 2$, $\Delta_2 = - M - 1$, $\Delta_3 = M$

Since $\Delta_2 = - M - 1$ is mini of Δ_1, Δ_3. So y_2 is incoming vector

To find out doing vector.

By mini ratio rule we find that y_{1a} is out going vector.

The key element is 1

Proceeding as before second simplex table

Second Table

		c_J	2	1	0	0	–M	w_B
B	c_B	w_B	Y_1	Y_2	Y_{1s}	Y_{2s}	Y_{1a}	Y_{i3}
x_2	1	1	–2	1	–1	0	1	–ive
x_{2s}	0	2	–1	0	(1)	1	1	2→
		0	0	–1	0	1 + M		
				↑	↓			

Proceeding as usual the third table as follows:

Third Table

		c_J	2	1	0	0	–M	w_B
B	c_B	w_B	Y_1	Y_2	Y_{1s}	Y_{2s}	Y_{1a}	Y_{i1}
x_2	1	3	–3	1	0	1	2	–ive
x_{1s}	0	2	–1	0	1	1	1	–ive
	$Z_J–c_J$		–5	0	0	1	2+M	
			↑					

Because all value of a are (–ve) therefore solution is no feasible solution.

Example 24:

Write the dual of L.P.P. and solve it.

Mini $Z = 8x_1 + 2x_2 + 4x_3$

s.t. $x_1 - 4x_2 - 2x_3 \geq 2;\ x_1 + x_2 - 3x_3 \geq -1$

$-3x_1 - x_2 + x_3 \geq 1;\ x_1, x_2, x_3 \geq 0.$

Solution:

The dual of this problem is given by

Max. $Z_D = 2w_1 - w_2 + w_3$

s.t. $w_1 + w_2 - 3w_3 \leq 8$

$-4w_1 + w_2 - w_3 \leq 2$

$-2w_1 - 3w_2 + w_3 \leq +4$

$w_1, w_2, w_3 \geq 0$

Introducing the slack variable w_{1s}, w_{2s} and w_{3s} respectively. The dual problem can be written as

Max $Z_D = 2w_1 - w_2 + w_1$

s.t. $w_1 + w_2 - 3w_3 + w_{1s} = 8$

$-4w_1 + w_2 - w_3 + w_{2s} = 2$

$-2w_1 - 3w_2 + w_3 + w_{3s} = +4$

$w_1, w_2, w_3, w_{1s}, w_{2s}, w_{3s} \geq 0$

Taking $w_1 = 0 = w_2 = w_3$ we have $w_{1s} = 8$,

$w_{2s} = 2\ w_{3s} = 4$,

which is the starting basic feasible solution of the dual.

First Table

		c_j	2	–1	1	0	0	0	w_B
B	c_B	w_B	Y_1	Y_2	Y_3	Y_{1s}	Y_{2s}	Y_{3s}	Y_{i1}
x_{1s}	0	8	1	1	–3	1	0	0	8
x_{2s}	0	2	–4	1	–1	0	1	0	–ive
x_{3s}	0	4	–2	–3	1	0	0	1	–ive
			–2	1	–1	0	0	0	
		↑			↓				

Since $\Delta_1 = -2$ is mini of $\Delta_1, \Delta_2, \Delta_3$. So y_1 is incoming vector and by mini ratio rule out going vector is y_{1s}.

The key element is $y_{11} = 1$

Proceeding as before the second simplex table

Second Table

		c_J	2	–1	1	0	0	0	w_B
B	c_B	w_B	Y_1	Y_2	Y_3	Y_{1s}	Y_{2s}	Y_{3s}	Y_{i3}
x_1	2	8	1	1	–3	1	0	0	–ive
x_{2s}	0	34	0	5	–13	4	1	0	–ive
x_{3s}	0	20	0	–1	–5	2	0	1	–ive
			0	3	–7	2	0	0	
					↑				

All w_B/y_{i3} is negative so solution is no feasible solution or unbounded solution.

Example 25:

Write the dual of the following problem

$$Min.\ Z = 3x_1 - 2x_2 + 4x_3$$

subject to $3x_1 + 5x_2 + 4x_3 \geq 7$

$$6x_1 + x_2 + 3x_3 \geq 4$$

$$7x_1 - 2x_2 - x_3 \leq 10$$

$$x_1 - 2x_2 + 5x_3 \geq 3$$

$$4x_1 + 7x_2 - 2x_3 \geq 2$$

$$x_1, x_2, x_3 \geq 0.$$

Solution:

First we shall convert the given L.P.P. into standard prima form.

Since the given problem is of minimization therefore all the constraints should have the sign $\geq$.

Thus the standard primal form of the given L.P.P. is

$$\text{Min } Z = 3x_1 - 2x_2 + 4x_3$$

subject to $3x_1 + 5x_2 + 4x_3 \geq 7$

$$6x_1 + x_2 + 3x_3 \geq 4$$

$$-7x_1 + 2x_2 + x_3 \geq -10$$

$$x_1 - 2x_2 + 5x_3 \geq 3$$

$4x_1 + 7x_2 - 2x_3 \geq 2,\ x_1,$

$x_2,\ x_3 \geq 0.$

The matrix form of the above problems

Min $Z = (3, -2, 4)\,(x_1, x_2, x_3)$

$$\text{subject to} \begin{bmatrix} 3 & 5 & 4 \\ 6 & 1 & 3 \\ -7 & 2 & 1 \\ 1 & -2 & 5 \\ 4 & 7 & -2 \end{bmatrix} \begin{bmatrix} x_1 \\ x_2 \\ x_3 \end{bmatrix} \geq \begin{bmatrix} 7 \\ 4 \\ -10 \\ 3 \\ 2 \end{bmatrix}.$$

Now the dual of the given primal is

Max. $Z_D = (7, 4, -10, 3, 2)\,(y_1, y_2, y_3, y_4, y_5)$

subject to

$$\begin{bmatrix} 3 & 6 & -7 & 1 & 4 \\ 5 & 1 & 2 & -2 & 7 \\ 4 & 3 & 1 & 5 & -2 \end{bmatrix} \begin{bmatrix} y_1 \\ y_2 \\ y_3 \\ y_4 \\ y_5 \end{bmatrix} \leq \begin{bmatrix} 3 \\ -2 \\ 4 \end{bmatrix}$$

or Max. $Z_D = 7y_1 + 4y_2 - 10y_3 + 3y_4 + 2y_5$

subject to $3y_1 + 6y_2 - 7y_3 + y_4 + 4y_5 \leq 3$

$5y_1 + y_2 + 2y_3 - 2y_4 + 7y_5 \leq 3$

$4y_1 + 3y_2 + y_3 + 5y_4 - 2y_5 \leq 4,$

$y_1, y_2, y_3, y_4, y_5 \geq 0.$

Example 26:

Use Dual problem to solve

Min. $Z = 3x_1 + x_2$

subject to $x_1 + x_2 \geq 1$

$2x_1 + 3x_2 \geq 2,$

$x_1, x_2 \geq 0.$

Solution:

The given L.P.P. is in standard primal form. The dual problem is given by

Max. $Z_D = w_1 + 2w_2$

subject to

$w_1 + 2w_2 \leq 3$

$w_1 + 3w_2 \le 1,\ w_1,\ w_2 \ge 0.$

Introducing slack variables w_3 and w_4 to change the constraint inequalities into equations, the dual problem becomes

Max. $Z_D = w_1 + 2w_2 + 0w_3 + 0w_4$

subject to

$$w_1 + 2w_2 + w_3 = 3$$
$$w_1 + 3w_2 + w_4 = 1,$$
$$w_1,\ w_2,\ w_3,\ w_4 \ge 0.$$

Initial B.F.S. is $w_3 = 3,\ w_4 = 1.$

Now we shall apply simplex method to obtain solution.

		c_j	1	2	0	0	Min ratio
B	c_B	w_B	W_1	W_2	W_3	W_4	w_B/W_2
W_3	0	3	1	2	1	0	3/2
W_4	0	1	1	(3)	0	1	<u>1/3</u>
$Z_D = 0$		Δ_j	1	2 ↑	0	0 ↓	w_B/W_1
W_3	0	7/3	1/3	0	1	–2/3	7
W_2	2	1/3	(1/3)	1	0	1/3	<u>1</u>
$Z_D = 2/3$		Δ_j	1/3 ↑	0 ↓	0	–2/3	
W_3	0	2	0	–1	1	–1	
W_1	1	1	1	3	0	1	
$Z_D = 1$	Δ_j	0	–1	0	–1		

Since all $\Delta_j \le 0$, therefore the dual problem has optimal solution

$$w_1 = 1,\ w_2 = 0,$$

Max. $Z_D = 1.$

Now the solution of the original primal problem form the last simplex table of the dual is

$$x_1 = -\Delta_3 = 0,$$
$$x_2 = -\Delta_4 = 1,$$

min. $Z = \text{max. } Z_D = 1.$

Thus, the optimal product mix is: none of the product A and 40 units of product B for a total conscription of Rs. 200.

Example 27:

Write the dual of the following linear programming problem and hence solve it.

$$Max.\ Z = 3x_1 - 2x_2$$

$$\text{subject to } x_1 \leq 4$$

$$x_2 \leq 6$$

$$x_1 + x_2 \leq 5$$

$$-x_2 \leq -1$$

$$x_1, x_2 \geq 0.$$

Solution:

The given problem is in standard primal form. The dual of the given primal is

$$\text{Min. } Z_D = 4w_1 + 6w_2 + 5w_3 - w_4$$

$$\text{subject to } w_1 + w_3 \geq 3$$

$$w_2 + w_3 - w_4 \geq -2,$$

$$w_1, w_2, w_3, w_4 \geq 0.$$

Changing the dual problem to maximization and introducing surplus variable w_5 and slack variable w_6 to change the inequalities into equations, the dual problem becomes

$$\text{Max.. } Z_D' = -4w_1 - 6w_2 - 5w_3 + w_4 + 0w_5 + 0w_6$$

$$\text{subject to } w_1 + w_3 - w_5 = 3$$

$$-w_2 - w_3 + w_4 + w_6 = 2,$$

$$w_1, w_2, \ldots, w_6 \geq 0.$$

Here, we have not introduced the artificial variable because in the first constraint equation w_1 will serve the purpose.

The starting B.F.S. is

$$w_1 = 3,\ w_6 = 2.$$

Now applying the simplex method to obtain the optimal solution of the dual problem

B	c_B	c_j / w_B	-4 / W_1	-6 / W_2	-5 / W_3	1 / W_4	0 / W_5	0 / W_6	Min. ratio w_B/W_4
W1	-4	3	1	0	1	0	-1	0	--
W_6	0	2	0	-1	-1	(1)	0	1	2
$Z_D' = -12$		Δ_j	0	-6	-1	1 ↑	-4	0 ↓	
W_1	-4	3	1	0	1	0	-1	0	
W_4	1	2	0	-1	-1	1	0	1	
$Z_D'=-10$		Δ_j	0	-5	0	0	-4	-1	

Since all Δ_j are negative or zero therefore the last table gives the optimal solution of the dual.

The optimal solution of the dual is

$w_1 = 3,\ w_2 = 0,\ w_3 = 0,\ w_4 = 2,$

Min. $Z_D = -$ Max. $Z_D' = 10$.

To read the solution of the primal from the final simplex table of the dual.

The optimal solution of the primal problem is

$x_1 = -\Delta_5 = 4,\ x_2 = -\Delta_6 = 1$ and Max. Z = Min. $Z_D = 10$.

Example 28:

Give the dual of the following problem and solve

Mini. $Z_P = x_1 + x_2,\ 2x_1 + x_2 \geq 4,\ x_1 + 7x_2 \geq 7,\ x_1, x_2 \geq 0.$

Solution:

The dual of the given problem is given by

Max. $Z_D = 4w_1 + 7w_2$

s.t. $2w_1 + w_2 \leq 1$

$w_1 + 7w_2 \leq 1$

$w_1, w_2 \geq 0$

Introducing the slack variable w_{1s} and w_{2s} respectively. The dual problem can be written as

Max. $Z_D = 4w_1 + 7w_2$

s.t. $2w_1 + w_2 + w_{1s} = 1$

$w_1 + 7w_2 + w_{2s} = 1$

$w_1, w_2, w_{1s}, w_{2s} \geq 0$

Taking $w_1 = 0 = w_2$, we have $w_{1s} = 1$, $w_{2s} = 1$, which is the starting basic feasible solution of the dual.

First Table

		c_J	4	7	0	0	w_B
B	c_B	w_B	Y_1	Y_2	Y_{1s}	Y_{2s}	y_{i2}
x_{1s}	0	4	2	1	1	0	4
x_{2s}	0	7	1	7	0	1	1 mini
		$Z_J - c_J$	–4	–7	0	0	

$\Delta_1 = -4$, $\Delta_2 = -7$

Since $\Delta_2 = -7$ is mini of Δ_1 and Δ_2 so y_2 is incoming vector and by minimum ratio rule outgoing vector is y_{2s}.

The key element is 7.

Proceeding as before the second simplex table.

Second Table

		c_J	4	7	0	0	w_B
B	c_B	w_B	Y_1	y_2	Y_{1s}	Y_{2s}	Y_{i1}
x_{1s}	0	3	13/7	0	1	–1/7	21/13 mini
x_2	7	1	1/7	1	0	1/7	7
		$Z_J - c_J$	–3	0	0	1	
			↑		↓		

Since $\Delta_1 = -3$ is mini so y_1 is incoming vector and by mini rule method outgoing vector is y_{1s}.

The key element is $y_{11} = 13/7$

Proceeding as before the third simplex table.

Third Table

		c_J	4	7	0	0
B	c_B	w_B	Y_1	Y_2	Y_{1s}	Y_{2s}
x_1	4	21/13	1	0	7/13	–1/13
x_2	7	10/13	0	1	–1/13	1/13
		$Z_J - c_J$	0	0	$\frac{21}{13}$	10/13

Since all $Z_j - c_j \geq 0$, so this solution of the dual problem is optimal solution of the dual is

$w_1 = 21/13$, $w_2 = 10/13$; Max $Z_D = \dfrac{31}{13}$ solution of the primal problem is $x_1 = \dfrac{21}{13}$, $x_2 = 10/13$; Mini. $Z_P = 31/13$.

Example 29:

Apply simplex method to solve the following problem

$$\text{Max. } Z = 30x_1 + 23x_2 + 29x_3$$

subject to $6x_1 = 5x_2 + 3x_3 \leq 26$

$$4x_1 + 2x_2 + 5x_3 \leq 7;\ x_1, x_2, x_3 \geq 0.$$

Hence or otherwise find the solution to the dual of the above problem.

Solution:

The given problem is of maximization. Introducing slack variables x_4, x_5 to change the constraint inequalities into equations, we get

$$\text{Max. } Z = 30x_1 + 23x_2 + 29x_3 + 0x_4 + 0x_5,$$

s.t. $6x_1 + 5x_2 + 3x_3 + x_4 = 26$

$$4x_1 + 2x_2 + 5x_3 + x_5 = 7,$$

$$x_1, x_2, \ldots, x_5 \geq 5.$$

The starting B.F.S. is

$x_1 = 0,\ x_2 = 0,\ x_3 = 0,$

$x_4 = 26,\ x_5 = 7.$

Now applying simplex method to obtain the optimal solution, we have

		c_j	30	23	29	0	0	Min. ratio
B	c_B	x_B	Y_1	Y_2	Y_3	Y_4	Y_5	x_B/Y_1
Y_4	0	26	6	5	3	1	0	26/6
Y_5	0	7	(4)	2	5	0	1	7/4
$Z=c_B$	$x_B=0$	Δ_j	30 ↑	23	29	0	0 ↓	x_B/Y_2
Y_4	0	31/2	0	2	−9/2	1	−3/2	31/4
Y_1	30	7/4	1	(1/2)	5/4	0	1/4	7/2
Z=105/2	Δ_j	0 ↓	8 ↑	−17/2	0	−15/2		

Y_4	0	17/2	–4	0	–19/2	1	–5/2
Y_2	23	7/2	2	1	5/2	0	1/2
Z=161/2	Δ_j	–16	0	–57/2	0	–23	/2

In the last simplex table all $\Delta_j \leq 0$ therefore the optimal solution is

$$x_1 = 0,\ x_2 = 7/2,\ x_3 = 0$$

and $\quad$ Max. $Z = 161/2$.

Dual Problem

The given problem is in standard primal form. The dual of the given problem is

$$\text{Min. } Z_D = 26w_1 + 7w_2$$

subject to $6w_1 + 4w_2 \geq 30$

$$5w_1 + 2w_2 \geq 23$$

$$3w_1 + 2w_2 \geq 29,$$

$$w_1, w_2 \geq 0.$$

To read the solution of the dual from the find simplex table of the primal problem.

$$w_1 = -\Delta_4 = 0,$$

$$w_2 = -\Delta_5 = 23/2$$

and $\quad$ min Z_D max $Z = 161/2$.

EXERCISES

1. Show that the value of the objective function of the dual for any feasible solution is never less than the value of the objective function of the primal corresponding to any feasible solution.
2. Show with the help of an example how when one solves an LP problem by simplex method going through infeasible but better than optimal solution, one directly goes through infeasible better than optimal solution of the dual LP problem. How is this fact utilized in the solution of the dual.
3. Define the dual of a linear programming problem. State the functional properties of duality.
4. Explain the primal-dual relationship.
5. Discuss in brief 'duality' in linear programming.

Use principle of Dual problem to solve the following problems:

6. One unit of product A contributes Rs. 7 and requires 3 units of raw material and 2 hours of labour. One unit of product B contributes Rs. 5 and requires one unit of raw material and one hour of labour. Availability of raw material at present is 48 units and there are 40 hours of labour,
   (a) Formulate this problem as a linear programming problem.
   (b) Write its dual.
   (c) Solve the dual by the simplex method and find the optimal product mix and the shadow prices of the raw material and labour.
7. What is dual simplex algorithm ? State various steps involved in the dual simplex algorithm.
8. What is the essential difference between regular simplex method and dual simplex method?
9. What is the principle of duality in linear programming? Explain its advantages.
10. What is duality? What is the significance of dual variables in a LP model?
11. State the general rules for formulating a dual LP problem from its primal.
12. Prove that if the primal has an unbounded solution, the dual has no solution and *vice-versa.*
13. Define the dual of given L.P.P. prove that dual of a dual is primal problem.
14. Define the dual of given L.P.P. prove that dual of a dual is primal problem.
15. A medical scientist claims to have found a cure for the common cold that consists of three drugs called K, S and H. His results indicate that the minimum daily adult dosage for effective treatment is 10 mg of drug K, 6 mg of drug S and 8 mg of drug H. Two substances are readily available for preparing pills for distribution to cold sufferers. Both substances contain all three of the required drugs. Each unit of substance A contains 6 mg, 1 mg and 2 mg of drugs K, S and H respectively and each unit of substance B contains 2 mg, 3 mg and 2 mg of the same drugs. Substance A costs Rs. 3 per unit and substance B costs Rs. 5 per unit.

(a) Find the least-cost combination of the two substances that will yield a pill designed to contain the minimum daily recommended adult dosage.

(b) Suppose that the costs of the two substances are interchanged so that substance A costs Rs. 5 per unit and substance B costs Rs. 3 per unit. Find the new optimal solution.

16. Write the dual of the given L.P.P. Also find the solution of the dual problem.

17. What is a shadow price? How does the concept relate to the dual of an LP problem?

18. State and prove the relationship between the feasible solutions of LP problem and its dual.

19. Five wagons are available at five stations 1, 2, 3, 4, and 5. These are required at five stations I, II, III, IV and V. The mileages between various stations are given by the following table:

20. A company produces three products: P, Q and R from three raw materials A, B and C. One unit of product P requires 2 units of A and 3 units of B. A unit of product Q requires 2 units of B and 5 units of C and one unit of product R requires 3 units of A, 2 units of B and 4 units of C. The company has 8 units of material A, 10 units of material B and 15 units of material C available to it. Profits per unit of products P, Q and R are Rs. 3, Rs. 5 and Rs. 4, respectively.

(a) Formulate this problem as an LP problem.

(b) How many units of each product should be produced to maximize profit?

(c) Write the dual of this problem.

21. Three food products are available at costs of Rs. 10, Rs. 36 and Rs. 24 per unit, respectively. They contain 1,000, 4,000 and 2,000 calories per unit, respectively and 200, 900 and 500 protein units per unit, respectively. It is required to find the minimum-cost diet containing at least 20,000 calories and 3,000 units of protein. Formulate and solve the given problem as a LP problem. Write the dual and use it to check the optimal solution of the given problem.

Find the dual of the following linear programming problems:

22. Max. $Z = 3x_1 + 2x_2$

subject to

$2x_1 + x_2 \leq 5,\ x_1 + x_2 \leq 3$

and $x_1, x_2 \geq 0$.

23. Max. $Z = 2x_1 + x_2$

subject to

$x_1 + 2x_2 \leq 10,\ x_1 + x_2 \leq 6$

$x_1 - x_2 \leq 2,\ x_1 - 2x_2 \leq 1$

and $x_1, x_2, x_3 \geq 0$.

24. Min. $Z = 2x_1 + 2x_2$

subject to

$2x_1 + 4x_2 \geq 1,$

$x_1 + 2x_2 \geq 1.$

$2x_1 + x_2 \geq 1,$

and $x_1, x_2 \geq 0$.

25. Max. $Z = 3x_1 + 2x_2$

subject to $x_1 + x_2 \geq 1,$

$x_1 + x_2 \leq 7,$

$x_1 + 2x_2 \leq 10,\ x_2 \leq 3$

and $x_1, x_2 \geq 0$.

26. Formulate the dual of the given L.P.P. and hence solve it:

Min. $Z = 3x_1 - 2x_2 + 4x_3$

subject to

$3x_1 + 5x_2 + 4x_3 \geq 7,$

$6x_1 + x_2 + 3x_3 \geq 4,$

$7x_1 - 2x_2 - x_3 \leq 10,$

$x_1 - 2x_2 + 5x_3 \geq 3$

$4x_1 + 7x_2 - 2x_3 \geq 2$

and $x_1, x_2, x_3 \geq 0$.

27. Solve the following problem by simplex method:

Max. $Z = 30x_1 + 23x_2 + 29x_3$

subject to

$6x_1 + 5x_2 + 3x_3 \leq 26,$

$4x_1 + 2x_2 + 5x_3 \leq 7,$

$x_1, x_2, x_3 \geq 0.$

Also read the solution to the dual of the above problem.

28. Use Dual problem theory to solve the given L.P.P.

    Min. $Z = 4x_1 + 3x_2 + 6x_3$

    subject to

    $x_1 + x_3 \geq 2,$

    $x_2 + x_3 \geq 5$

    and $x_1, x_2, x_3 \geq 0.$

29. Solve the following L.P.P. using Dual problem theory

    Min. $Z = 50x_1 - 80x_2 - 140x_3,$

    subject to

    $x_1 - x_2 - 3x_3 \geq 4,$

    $x_1 - 2x_2 - 2x_3 \geq 3$

    and $x_1, x_2, x_3 \geq 0.$

30. Apply the principle of Dual problem to solve the following L.P.P.

    Max. $Z = 3x_1 - 2x_2$

    subject to

    $x_1 + x_2 \leq 5,$

    $x_1 \leq 4,\ 1 \leq x_2 \leq 6$

    and $x_1, x_2 \geq 0.$

31. Use Dual problem theory to solve

    Min. $Z = 3x_1 + x_2$

    subject to $x_1 + x_2 \geq 1,$

    $2x_1 + 3x_2 \geq 2,$

    $x_1, x_2, x_3 \geq 0.$

32. Find the dual of the following L.P.P.

    Max. $Z = 2x_1 - x_2$

    subject to

    $x_1 + x_2 \leq 10,$

    $-2x_1 + x_2 \leq 2$

    $4x_1 + 3x_2 \geq 12,$

    $x_1, x_2, \geq 0.$

Solve the primal problem by simplex method and deduce from it the solution to the dual problem.

33. Find the dual of the following problem

Min. $Z = 6x + 5y + 2x$

subject to $x + 3y + 2z \geq 5,$

$2x + 2y + z \geq 2,$

$4x - 2y + 3z \geq -1,$

$x, y, z \geq 0.$

Hence or otherwise solve the primal problem.

34. Use Dual problem to solve the problem

Min. $Z = 10x_1 + 6x_2 + 2x_3$

subject to

$-x_1 + x_2 + x_3 \geq 1,$

$3x_1 + x_2 - x_3 \geq 2,$

and $x_1, x_2, x_3 \geq 0.$

35. Write the dual of the following problem

Max. $Z = 2x_1 + 3x_2,$

subject to

$2x_1 + 2x_2 \leq 10,$

$2x_1 + x_2 \leq 6,$

$x_1 + 2x_2 \leq 6,$

$x_1, x_2, \geq 0.$

Solve the primal and then find the solution to the dual.

36. Max. $Z = x_1 - x_2 + 3x_3$

subject to

$x_1 + x_2 + x_3 \leq 10$

$2x_1 - x_3 \leq 2$

$2x_1 - 2x_2 + 3x_3 \leq 6$

$x_1, x_2, x_3 \geq 0.$

37. Max. $Z = 3x_1 + 5x_2 + 4x_3$

subject to

$2x_1 + 3x_2 \leq 8$

$2x_2 + 5x_3 \le 10$

$3x_1 + 2x_2 + 4x_3 \le 15$

$x_1, x_2, x_3 \ge 0.$

38. Max. $Z = x_1 + 3x_2$

subject to

$3x_1 + 2x_2 \le 6$

$3x_1 + x_2 = 4,$

$x_1, x_2 \ge 0.$

39. Min. $Z = x_1 + x_2 + x_3$

subject to

$x_1 - 3x_2 + 4x_3 = 5$

$x_1 - 2x_2 \le 3$

$2x_2 - x_3 \ge 4.$

$x_1, x_2 \ge 0$, x_2 is unrestricted.

40. Min. $Z = 7x_1 + 3x_2 + 8x_3$

subject to

$8x_1 + 2x_2 + x_3 \ge 3$

$3x_1 + 6x_2 + 4x_3 \ge 4$

$4x_1 + x_2 + 5x_3 \ge 1$

$x_1 + 5x_2 + 2x_3 \ge 7,$

$x_1, x_2, x_3 \ge 0.$

41. Max. $Z = 3x_1 + x_2 + x_3 - x_4$

subject to

$x_1 + 5x_2 + 3x_3 + 4x_4 \le 5$

$x_1 + x_3 = -1$

$x_3 - x_4 \ge -5,$

$x_1, x_2, x_3, x_4 \ge 0.$

42. Max. $Z = 3x_1 + x_2 + 4x_3 + x_4 + 9x_5$

subject to

$4x_1 - 5x_2 - 9x_3 + x_4 - 2x_5 \le 6$

$2x_1 + 3x_2 + 4x_3 - 5x_4 + x_5 \le 9$

$x_1 + x_2 - 5x_3 - 7x_4 + 11x_5 \le 10,$

43. A company wishes to get at least 160 million 'audience exposures' everytime one of the advertisements is seen or heard by a person. Because of the nature of the product, the company wants at least 60 million of these exposures to involve persons with family income of over Rs 10,000 a year and at least 80 million of the exposures to involve person between 18 and 40 years of age. The relevant information pertaining to the two advertising media under considerationumagazine and television, is given below.

	Magazine	*Television*
Cost per advertisement (Rs thousand)	40	200
Audience per advertisement (million)	4	40
Audience per advertisement with income over Rs 10,000 (million)	3	10
Audience (per advertisement) in the age group 18-40 (million)	8	10

The company wishes to determine the number of advertisements each to be released in magazines and television so as to keep the advertisement expenditure to the minimum.

Formulate this problem as a linear programming problem. What will be the minimum expenditure and its allocation among the two media? Write the 'dual' of this problem. You may solve the 'dual' problem to find answer to the problem on hand.

44. A diet conscious housewife wishes to ensure certain minimum intake of vitamins A, B and C for the family. The minimum daily (quantity) needs of the vitamins A, B and C for the family are respectively 30, 20 and 16 units. For the supply of these minimum vitamin requirements, the housewife relies on two fresh foods I and II. The first one provides 7, 5 and 2 units of the three vitamins per gram, respectively and the second one provides 2, 4 and 8 units of the same three vitamins per gram of the foodstuff, respectively. The first foodstuff costs Rs. 3 per gram and the second Rs. 2 per gram. The problem is how many grams of each foodstuff should the housewife buy everyday to keep her food bill as low as possible?

    (a) Formulate this problem as an LP model.

    (b) Write and then solve the dual problem.

45. Prove that the necessary and sufficient condition for any LP problem and its dual to have optimal solutions is that both have feasible solutions.

46. How the concept of duality can be useful in managerial decision-making?

47. Find the solution of the following problem by simplex method:

$$\text{Max. } Z_x = 40x_1 + 50x_2$$

$$\text{subject to } 2x_1 + 3x_2 \leq 3,\ 8x_1 + 4x_2 \leq 5,\ x_1, x_2 \geq 0.$$

48. If (x_{ij}), $i = 1, 2, \ldots, n$; $j = 1, 2, \ldots, n$ is an optimal solution for an assignment problem with cost (c_{ij}), then it is also optimal for the problem with cost (c_{ij}') when

$$c_{ij}' = c_{ij} \text{ for } i, j = 1, 2, \ldots, n;\ j \neq k$$

$$c_{ik}' = c_{ik} - A, \text{ where A is a constant.}$$

49. State the duality theorem for an LP problem and hence develop the computational algorithm for solving it by dual simplex method.

50. Show with the help of an example how when one solves an LP problem by simplex method going through infeasible but better than optimal solution, one indirectly goes through infeasible but better than optimal solution of the dual LP problem. How this fact is utilized in the solution of the dual?

4

Variation of Analysis Problems

INTRODUCTION

Analysis problem and parametric linear programming are the two techniques that evaluate the relationship between the optimal solution and changes in the LP model parameters. Analysis *problem considered the effects of variations in the input coefficients (also called parameters) when these coefficients are changed one at a time, whereas parametric analysis considered the effects of simultaneous changes in data, where the coefficients change as a function of one parameter.*

However, in real-world situations some data may change over time because of the dynamic nature of the business: Such changes in any of these parameters may raise doubt on the validity of the optimal solution of the given LP model. Thus, a decision-maker in such situations would like to know how sensitive the optimal solution is to the changes in the original input data values.

While sensitivity analysis provides the sensitive ranges (both lower and upper limits) within which the LP model parameters can vary without changing the optimality of the current optimal solution; parametric linear programming provides information to such changes outside the sensitive range, as well as to changes in more than one parameter at a time.

The problems of the type discussed above arise in general, when slight changes are made in the parameters or the structure of a given L.P.P. after its optimum solution has been attain. An analysis of such post optimal problem can thus, be formed. Post optimality analysis the changes (variations) in an L.P.P. which are usually studied by post optometry analysis indeed:

1. Changes in the requirement vector b.
2. Changes in the cost vector c these may further be classified as the changes in :
   (a) The cost coefficients associated with the basic variables, and.
   (b) The cost coefficients associated with the non-basic variables.

3. Changes in the elements of coefficient matrix A, these may further be classified as the changes in.
   (a) Those column vectors of A which form a basis set, and
   (b) Those column vectors of 'A' which do not form a basis set.
4. Structural changes due to the addition and deletion of some linear constraints.
5. Structural changes due to the addition and subtraction (deletion) of same variables.

The process of studying the sensitivity of the optimal solution of an LP problem is also called *post-optimality analysis* because it is done after an optimal solution, assuming a given set of parameters, has been obtained for the model.

VARIATION IN THE VECTOR b REQUIREMENT

We know that the condition of optimality for the B.F.S. of a L.P.P. is $\Delta_j = c_j - Z_j \leq 0$.

Since Δ_j does not involve any of b_i if any component b_i of the requirement vector $b = [b_1, b_2, ..., b_m]$ is changed will not affect the conditions of optimality. Hence if any component b_i is changed to $b_i + \Delta b_i$ then the new solution thus obtained will remain optimal. But $x_B = B^{-1} b$ depends on b, therefore any change in b may affect the feasibility of the optimal solution *i.e.*, the optimal solution obtained by changing b may or may not be feasible.

Thus, a change in b_i must be of the magnitude vector b be changed to $b_j + \Delta b_j$, therefore if the new requirement vector is b*, then

$$b^* = [b_1, b_2, ... ,b_l + \Delta b_l, ... ,b_m].$$

If x_B^* is the solution of the new L.P.P. obtained by changing b_l to $b_t + Db_l$, then

$$x_B^* = B^{-1} b^* \text{ where B is the optimal basis.}$$

Let $B^{-1} = (\beta_1, \beta_2, ... ,\beta_m)$

$$= \begin{bmatrix} \beta_{11} & \beta_{12} & \cdots & \beta_{1t} & \cdots & \beta_{1m} \\ \beta_{21} & \beta_{22} & \cdots & \beta_{2t} & \cdots & \beta_{2m} \\ \vdots & \vdots & & \vdots & & \vdots \\ \beta_{i1} & \beta_{i2} & \cdots & \beta_{il} & \cdots & \beta_{im} \\ \vdots & \vdots & & \vdots & & \vdots \\ \beta_{m1} & \beta_{m2} & \cdots & \beta_{ml} & \cdots & \beta_{mm} \end{bmatrix}$$

l-th component

Since $b^* = [b_1, b_2, \ldots, b_l + \Delta b_l, \ldots, b_m]$

$= [b_1 + 0, b_2 + 0, \ldots, b_l + \Delta\, b_l, \ldots, b_m + 0]$

$= [b_1, b_2, \ldots, b_l, \ldots, b_m] + [0, 0, \ldots, \Delta\, b_l, \ldots, 0]$

$= b + [0, 0, \ldots, \Delta\, b_l, \ldots, 0]$

$\therefore x_B^* = B^{-1}.\ b^*$

$= B^{-1}.\ \{b + [0, 0, \ldots, \Delta b_i, \ldots, 0]\}$

$= B^{-1}\ b + B^{-1}.\ [0, 0, \ldots, \Delta b_i, \ldots, 0]$

l-th component

$= x_B + [\beta_1, \beta_2, \ldots, \beta_l, \ldots, \beta_m].\ [0, 0, \ldots, \Delta b_l, \ldots, 0]$

$= x_B + \beta_l.\ \Delta b_l$

$= [x_{B1}, x_{B2}, \ldots, x_{Bl}, \ldots, x_{Bm}] + [\beta_{1l}, \beta_{2l}, \ldots, \beta_{ll}, \ldots, \beta_{ml}]\ .\ \Delta b_l.$

$= [x_{B1} + \beta_{1l}\ \Delta b_l, \ldots, x_{Bl} + \beta_{ll}\ \Delta b_l, \ldots, x_{Bm} + \beta_{ml}.\ \Delta b_l]$

Now if the solution x_B^* is feasible, then

$x_{Bi} + \beta_{il}\ \Delta b_l \geq 0$ for all $i = 1, 2, \ldots, m$

$\Rightarrow \beta_{il}\ \Delta b_l \geq -\ x_{Bi}$

$\therefore \Delta b_l \geq -\ \dfrac{x_{Bi}}{\beta_{it}}$, for $\beta_{il} > 0$

and $\quad \Delta b_l \leq -\ \dfrac{x_{Bi}}{\beta_{il}}$. for $\beta_{il} < 0$.

Hence the range of Δ_l so that the optimal solution $\mathbf{x}_B^*$ also remains feasible is given by

$$\underset{\beta_{il} > 0}{\text{Max}}\left[\frac{x_{Bi}}{\beta_{il}}\right] \leq \Delta b_l \leq \underset{\beta_{il} < 0}{\text{Min.}}\left[-\frac{x_{Bi}}{\beta_{il}}\right]$$

and the new value of the optimal solution is given by (1).

To find the change in the value of the objective function.

The given value of the objection function for a requirement vector is given by

$$Z = c_B\ x_B = \sum_{i=1}^{m} c_{Bi} \cdot x_{Bi}\ .$$

When b_l is changed to $b_l + \Delta b_l^*$ if Z^* is the value of the objective function, then $Z^* = c_B\ x_B$

$$= (c_{B1}, c_{B2}, \ldots, c_{Bm}) . [x_{B1} + b_{1l} \Delta_{bl}, \ldots, x_{Bm} + b_{ml} \Delta b_l]$$

$$= \sum_{i=1}^{m} c_{Bi} . (x_{Bi} + \beta_{il} . \Delta b \ )$$

$$= \sum_{i=1}^{m} c_{Bi} . x_{Bi} + \sum_{i=1}^{m} x_{Bi} . \beta_{il} . \Delta b_1$$

$$= Z + \sum_{i=1}^{m} x_{Bi} . \beta_{il} . \Delta b_1$$

Hence if Δb_l (change in b_l) satisfies (2), then the solution x_B given by (1) is also optimal feasible solution. and the value of the objective function is improved by an amount*

$$\sum_{i=1}^{m} x_{Bi} . \beta_{il} . \Delta b_1 .$$

VARIATION IN THE PRICE VECTOR c

Consider the L.P.P Max. $Z = cx$, s.t. $Ax = b$, $x \geq 0$. If x_B is the optimal basic feasible solution and B the optimal basis matrix, we have

$$x_B = B^{-1} b.$$

It is clear that x_B is independent of therefore the change in some component c_j of c will not change x_B, i.e., x_B will always remain basic feasible solution.

Condition of optimality for the solution x_B, is $\Delta_j = c_j - Z_j \leq 0$ for all j not in the basis which is satisfied before any change in c_j. But when c_j changes, the condition of optometry may not be satisfied. The change in the price vector c maybe made in the following two ways:

(i) Variation in $c_j \notin c_B$

(i.e., Change in c_j Which is the Price of the Non-Basic Variable x_j)

Since x_B is an optimal solution of the L.P.P. (maximization problem)

$\therefore \Delta_j = c_j - Z_j \leq 0$, for all j not in the basis.

If Δc_k is the change in the cost c_k and c_k is not present in c_B (the vector of the costs associated to basic variables) then c_B remains unaltered with this change. Also there is no change in B.

$\therefore Z_j = c_B B^{-1} \alpha_j = c_B Y_j$ also remains unaltered.

If x_B is still an optimal solution of the given L.P.P. when c_k changes to $c_k + \Delta c_k$ then, $c_j - Z_j$ remains same for all j ($\neq$ k) not in the basis,

$\therefore$ We must have

$$(c_k + \Delta c_k) - Z_k \leq 0$$

or $\quad \Delta c_k \leq Z_k - c_k \ (= -\Delta_k)$. ...(1)

Hence if $c_k \notin c_B$ changes to $c_k + \Delta c_k$ such that $\Delta c_k \leq Z_k - c_k$ $(= -\Delta_k)$, the value of the objective function and the optimal solution of the problem remains unchanged. Note that there is no lower bound to Δc_k.

(ii) Variation in $c_j \in c_B$.

(i.e., Change in c_j Which is the Price of the Basic Variable)

We know that

$$Z_j = c_B B^{-1} a_j = c_B Y_j = \sum_{i=1}^{m} c_{Bi} Y_{ij}.$$

Let Δc_{Bk} be the change in c_{Bk}, the price of the basic variable x_{Bk} then if Z_j^* is the value of Z_j in this solution, then we have

$$Z_j^* = \sum_{\substack{i=1 \\ i \neq k}}^{m} c_{Bi} y_{ij} + (c_{Bk} + \Delta c_{Bk}) y_{kj}$$

$$= \sum_{i=1}^{m} c_{Bi} y_{ij} + y_{kj} \Delta c_{Bk} = Z_j + y_{kj} \Delta c_{Bk}$$

$\therefore c_j - Z_j^* = c_j - (Z_j + y_{kj} \Delta c_{Bk}) = (c_j - Z_j) - y_{kj} \Delta c_{Bk}$.

The solution $\mathbf{x}_B$ will remain optimal for the change Δc_{Bk} in c_{Bk} if

$c_j - Z_j^* \leq 0$ for all j not in the basis

or $(c_j - Z_j) - y_{kj} \Delta c_{Bk} \leq 0$

or $y_{kj}, \Delta c_{Bk} \geq c_j - Z_j$

$$\therefore \Delta c_{Bk} \geq \frac{c_j - Z_j}{y_{kj}}, \text{ for } y_{kj} > 0$$

and $$\Delta c_{Bk} \leq \frac{c_j - Z_j}{y_{kj}}, \text{ for } y_{kj} < 0. \qquad ...(2)$$

Here the range of Δc_{Bk} (change in the price of c_{Bk} corresponding tooth basis variable x_{Bk}) so that the solution remains optimal is given by

$$\underset{y_{kj} > 0}{\text{Max}} \cdot \left[\frac{c_j - Z_j}{y_{kj}}\right] \leq \Delta c_{Bk} \leq \underset{y_{kj} < 0}{\text{Min}} \cdot \left[\frac{c_j - Z_j}{y_{kj}}\right] \qquad ...(3)$$

for all j corresponding to which α_j is not in the optimal basis.

If no $y_{kj} > 0$, there is no lower bound to Δc_{Bk} and if no $y_{kj} < 0$, there is no upper bound to Δc_{Bk}.

To find the change in the value of the objective function.

The value of the objective function for the price vector c is given by

$$Z = c_B . x_B = \sum_{i=1}^{m} c_{Bi} . x_{Bi} .$$

When $c_{Bk} \in c_B$ is changed to $c_{Bk} + \Delta c_{Bk}$, if Z^* is the value of the objective function then

$$Z^* = \sum_{\substack{i=1 \\ i \neq k}}^{m} c_{Bi} . x_{Bi} + (c_{Bk} + \Delta c_{Bk}) x_{Bk}$$

$$= \sum_{i=1}^{m} c_{Bi} . x_{Bi} + x_{Bk} . \Delta c_{Bk} = Z + x_{Bk} \Delta c_{Bk}.$$

Hence if Δc_{Bk} (change in c_{Bk}) satisfies (3), the solution x_B will remain optimal and the value of the objective function will be improved by an amount $x_{Bk} . \Delta c_{Bk}$ where x_{Bk} is the basic variable corresponding to c_{Bk}.

THE COEFFICIENT MATRIX a VARIATION IN THE COMPONENT a_{ij}

Let the element a_{lk} of the *l*-th row and k-th column of the coefficient matrix A be changed to $a_{lk} + \Delta a_{lk}$. Now there are two possibilities according as a_{lk} is or not the element of the optimal basis B.

Case I. When a_{lk} is not an Element of Optimal Basis B

If a_{lk} is not an element of the optimal basis B, a change in a_{lk} will not affect B and so B^{-1} remains the same. Thus the change in such a_{lk} will not affect the solution $x_B = B^{-1}b$. Hence the feasibility of the solution x_B is not affected. But the change of a_{lk} may change the optimality of the solution *i.e.*, the solution x_B may not be the optimal solution of the new L.P.P. (obtained by changing a_{lk} to $a_{lk} + \Delta a_{lk}$). Thus we have to find the range of variation of a_{lk} so that the solution still remains optimal.

Since all the column voters except a_k of the matrix A remain unaffected by the variation Δa_{lk} in a_{lk}.

$\therefore \Delta_j = c_j - Z_j \leq 0$ for all j ($\neq$ k) not in the basis.

Thus the solution x_B will remain optimal for the new L.P.P. also, if

$$\Delta_k^* = c_k - Z_k^* \leq 0 \qquad ...(1)$$

where Δ_k^* and Z_k^* are Δ_k and Δ_k for the new L.P.P.

Let $\quad B^{-1} = (\beta_1, \beta_2, ... ,\beta_l, ... ,\beta_m)$

and $\quad \alpha_k = [a_{lk}, a_{2k}, ..., a_{lk}, ... , a_{mk}]$

If α_k^* is α_k for the new L.P.P. then

$$\alpha_k^* = [a_{lk}, a_{2k}, \ldots , a_{lk} + \Delta a_{lk}, \ldots, a_{mk}]$$
$$= [a_{1k} + 0, a_{2k} + 0, \ldots ,a_{lk} + \Delta_{lk}, \ldots ,a_{mk} + 0]$$
$$= [a_{1k}, a_{2k}, \ldots ,a_{lk}, \ldots ,a_{mk}] + [0, 0, \ldots ,\Delta a_{lk}, \ldots ,0]$$
$$= \alpha_k + [0, 0, \ldots ,\Delta a_{lk}, \ldots ,0]$$

l-th component

and $\quad Z_k^* = c_B\, B^{-1}\, \alpha_k^*$

$$= c_B.\, B^{-1}\, \{\alpha_k + [0, 0, \ldots ,\Delta a^{lk}, \ldots 0]\}$$
$$= c_B.B^{-1}\, \alpha_k + c_B.B^{-1}\, [0, 0, \ldots ,\Delta a_{lk}, \ldots ,0]$$
$$= Z_k + c_B.(\beta_1, \beta_2, \ldots ,\beta_l, \ldots ,\beta_m).[0, 0, \ldots \Delta a_{lk}, \ldots ,0]$$
$$= Z_k + c_B\, (\beta_l\, \Delta a_{lk})$$
$$= Z_k + \Delta a_{lk}.\, c_B\, \beta_l$$
$$= Z_k + \Delta a_{lk}\, (c_{B1}, c_{B2}, \ldots ,c_{Bl}, \ldots ,c_{Bm}) \times [\beta_{1l}, \beta_{2l}, \ldots ,\beta^{ll}, \ldots \beta_{ml}]$$
$$= Z_k + \Delta a^{lk} \sum_{i=1}^{m} c_{Bi} . \beta_{il}\ .$$

$\therefore$ From (1), the solution x_B will remain optimal if

$$\Delta^{k*} = c_k - Z_k^* = c_k - Z_k - \Delta a_{lk} \sum_{i=1}^{m} c_{Bi} . \beta_{il} \le 0$$

$$\text{or } \Delta a^{lk} \sum_{i=1}^{m} c_{Bi} . \beta_{il} \ {}^3\ c^k - Z^k\ (= \Delta_k)$$

$$\therefore \Delta a_{lk} \ge \frac{\Delta_k}{\sum_{i=1}^{m} c_{Bi}\beta_{il}} \quad \text{for } \sum_{i=1}^{m} c_{Bi}\beta_{il} > 0.$$

$$\text{and} \qquad \Delta a_{lk} \le \frac{\Delta_k}{\sum_{i=1}^{m} c_{Bi}\beta_{il}} \quad \text{for } \sum_{i=1}^{m} c_{Bi}\beta_{il} < 0.$$

Hence the range of Δa_{lk} (change in $a_{lk} \notin B$), so that the solution x_B remains optimal and feasible, is given by

$$\left[\frac{\Delta_k}{\left(\sum_{i=1}^{m} c_{Bi}\beta_{il}\right) > 0}\right] \le \Delta a_{lk} \le \left[\frac{\Delta_k}{\left(\sum_{i=1}^{m} c_{Bi}\beta_{il}\right) < 0}\right] \qquad \ldots(2)$$

If $\sum_{i=1}^{m} c_{Bi} . \beta_{il} = 0$, $\Delta\, a_{lk}$ is unrestricted.

If $\sum_{i=1}^{m} c_{Bi} \cdot \beta_{il} > 0$, there is no upper bound to $\Delta\, a_{lk}$, and if $\sum_{i=1}^{m} c_{Bi} \cdot \beta_{il} < 0$, there is no lower bound to $\Delta\, a_{lk}$.

Note: There is no change in the value of the objective function if D a_{lk} satisfies (2).

Case II. When a_{lk} is an Element of the Optimal Basis B

Since a_{lk} is an element of the optimal basis B then if a_{lk} is changed to $a_{lk} + \Delta\, a_{lk}$, the optimal basis B will certainly be changed and hence B^{-1} will also be changed. Consequently x_B = B–1 b and Zj = $c_B B^{-1} \alpha_j$ will also change. A change in Z_j may disturb the optimality condition $\Delta_j = c_j - Z_j \leq 0$ while a change in x_B may disturb the feasibility of the solution. *Hence our aim is to find the range of variation of a_{lk} so that neither the optimality nor the feasibility of the solution is disturbed.*

When the element $a_{lk} \in B$ is changed to $a_{lk} + \Delta a_{lk}$ then let the optimal basis B, the solution x_B and Z_j be represented by B*, x_B* and Z^*_j respectively.

Let $\quad B = (b_1, b_2, \ldots, b_m)$ and α_k = bp

$\therefore a_{lk} = b_{lp}$ and $a_{lk} + \Delta\, a_{lk} = b_{lp} + \Delta\, b_{lp}$

$$\therefore B^* = B = \begin{pmatrix} 0 & 0 & \ldots & 0 & \ldots & 0 \\ 0 & 0 & \ldots & 0 & \ldots & 0 \\ \vdots & \vdots & & \vdots & & \vdots \\ 0 & 0 & \ldots & \Delta b_{lp} & \ldots & 0 \\ \vdots & \vdots & & \vdots & & \vdots \\ 0 & \ldots & 0 & \ldots & 0 & 0 \end{pmatrix} \leftarrow l - \text{th row}.$$

p-th column

$= B + \Delta\, b_{lp}.\, 0_{lp}$, where 0_{lp} is the null matrix except for the (l, p)th element which equals to unity

$= B\,(I + B^{-1}\,\Delta\, b_{lp}.\, 0_{lp})$

$\therefore B^{*-1} = \{B(I + B^{-1}\,\Delta b_{lp}.\, 0_{lp})\}-1 = (I + B^{-1}\,\Delta b_{lp}.\, 0_{lp})^{-1}.\, B^{-1}$ ...(3)

$[\because (AB)^{-1} = B^{-1}\, A^{-1}]$

If $B^{-1} = (\beta_1, \beta_2, \ldots, \beta_m)$,

$$I + B^{-1}\,\Delta b_{lp}\,.0_{lp} = I + B^{-1} \begin{pmatrix} 0 & 0 & \ldots & 0 & \ldots & 0 \\ 0 & 0 & \ldots & 0 & \ldots & 0 \\ \vdots & \vdots & & \vdots & & \vdots \\ 0 & 0 & \ldots & \Delta b_{lp} & \ldots & 0 \\ \vdots & \vdots & & \vdots & & \vdots \\ 0 & \ldots & 0 & \ldots & 0 & 0 \end{pmatrix}$$

$$= I + \xrightarrow{\text{p-th row}} \begin{bmatrix} \beta_{11} & \beta_{12} & \cdots & \beta_{1l} & \cdots & \beta_{1m} \\ \beta_{21} & \beta_{22} & \cdots & \beta_{21} & \cdots & \beta_{2m} \\ \vdots & \vdots & & \vdots & & \vdots \\ \beta_{pl} & \beta_{pl} & \cdots & \beta_{pl} & \cdots & \beta pm \\ \vdots & \vdots & & \vdots & & \vdots \\ \beta_{ml} & \beta_{m2} & \cdots & \beta_{ml} & \cdots & \beta_{mm} \end{bmatrix}$$

l-th column

$$\times \begin{pmatrix} 0 & 0 & \cdots & 0 & \cdots & 0 \\ 0 & 0 & \cdots & 0 & \cdots & 0 \\ \vdots & \vdots & & \vdots & & \vdots \\ 0 & 0 & \cdots & \Delta b_{lp} & \cdots & 0 \\ \vdots & \vdots & & \vdots & & \vdots \\ 0 & \cdots & 0 & \cdots & 0 & 0 \end{pmatrix} \leftarrow \text{l-th row}$$

p-th column

$$= \xrightarrow{\text{p-th row}} \begin{pmatrix} 1 & 0 & \cdots & 0 & \cdots & 0 \\ 0 & 1 & \cdots & 0 & \cdots & 0 \\ \vdots & \vdots & & \vdots & & \vdots \\ 0 & 0 & \cdots & 1 & \cdots & 0 \\ \vdots & \vdots & & \vdots & & \vdots \\ 0 & 0 & \cdots & 0 & \cdots & 0 \end{pmatrix}$$

p-th column

$$+ \begin{pmatrix} 0 & 0 & \cdots & \beta_{1l}\Delta b_{lp} & \cdots & 0 \\ 0 & 0 & \cdots & \beta_{2l}\Delta b_{lp} & \cdots & 0 \\ \vdots & \vdots & & \vdots & & \vdots \\ 0 & 0 & \cdots & \beta_{pl}\Delta b_{lp} & \cdots & 0 \\ \vdots & \vdots & & \vdots & & \vdots \\ 0 & 0 & \cdots & \beta_{ml}\Delta b_{lp} & \cdots & 0 \end{pmatrix} \leftarrow \text{p-th row}$$

p-th column

$$\Rightarrow x_B{}^* = \begin{bmatrix} x_{B1}\, 0 \dfrac{\beta_{1l}.\Delta b_{lp}}{D}.x_{Bp} \\ x_{B2} - \dfrac{\beta_{2l}.\Delta b_{lp}}{D}.x_{Bp} \\ \vdots \\ \dfrac{1}{D} x_{Bp} \\ x_{Bm} - \dfrac{\beta_{ml}.\Delta b_{lp}}{D}.x_{Bp} \end{bmatrix} \qquad \ldots(4)$$

The solution $\mathbf{x}_B^*$ is feasible if $\mathbf{x}_B^* \geq 0$.

$\therefore$ From (4), the solution x_B^* is feasible,

$$\text{if } x_{Bi} - \frac{\beta_{il}\,.\,\Delta b_{lp}}{D}\,.\,x_{Bp} \leq 0 \qquad \text{...(5)}$$

for all i = 1, 2, ... ,m (i ≠ p)

$$\text{and} \qquad \frac{1}{D}\,.\,x_{Bp} \geq 0. \qquad \text{...(6)}$$

Since $x_{Bp} \geq 0$

$\therefore$ (6) will hold of we have

$\therefore$ from (5), (assuming that (6) is true), we have

$x_{Bi}\, D - \beta_{il}.\, \Delta b_{lp}\, x_{Bp} \geq 0$

$\Rightarrow x_{Bi}.\, (1 + \beta_{pl}\, \Delta b_{lp}) - \beta_{il}\, \Delta b_{pl}\, x_{Bp} \geq 0$

$\Rightarrow x_{Bi} \geq (\beta_{il}\, x_{Bp} - \beta_{pl}\, x_{Bi}).\, \Delta b_{lp}$ for all i ≠ p

$$\therefore \Delta\, b_{lp} \leq \frac{x_{Bi}}{\beta_{il} x_{Bp} - \beta_{pl} x_{Bi}}, \text{ for } \beta_{il}\, x_{Bp} - \beta_{pl}\, x_{Bi} < 0$$

$$\text{and} \qquad \Delta\, b_{lp} \geq \frac{x_{Bi}}{\beta_{il} x_{Bp} - \beta_{pl} x_{Bi}}, \text{ for } \beta_{il}\, x_{Bp} - \beta_{pl}\, x_{Bi} < 0.$$

Hence the range of $\Delta\, a_{lk}$ $(= \Delta\, b_{lp}) \in B$, so that the solution remains feasible, is given by

$$\text{Max.} \left[\frac{x_{Bi}}{P_i < 0}\right] \leq \Delta a_{lk} \leq \text{Min} \left[\frac{x_{Bi}}{P_i > 0}\right] \qquad \text{...(7)}$$

where $\qquad D = 1 + \beta_{pl}\, \Delta\, \beta_{lp}, = 1 + \beta_{pl}\, \Delta\, a_{lk} > 0.$

$\qquad Pi = \beta_{il}\, x_{Bp} - \beta_{pl}\, x_{Bi}.$

If no $P_i < 0$, there is no lower bound to $\Delta\, a_{lk}$ and if no $P_i > 0$, there is no upper bound to $\Delta\, a_{lk}$.

To Find the Range of Variation of a_{lk} for the Optimality of the Solution

For optimality of the solution x_B^*, we must have

$\Delta_j^* = c_j - Z_j^* \leq 0$

For all j not in the basis.

$Z_j^* = c_B\ B^{*-1}\ \alpha_j = c_B\ (I + B^{-1}.\ \Delta\ b_{lp}\ 0_{lp})^{-1}.\ B^{-1}\ \alpha_j$

$= c_B\ (I + B^{-1}.\ \Delta\ b_{lp}\ 0_{lp})^{-1}.\ Y_j$

Since $B^{-1}\ \alpha_j = Y_j = [y_{1j}, y_{2j}, \dots, y_{pj}, \dots, y_{mj}]$.

Using (4) and simplifying as before as in (1), we have

$$Z_j^* = c_B \begin{bmatrix} y_{1j} - \dfrac{\beta_{1l}\Delta b_{lp}}{D}.y_{pj} \\ y_{2j} - \dfrac{\beta_{2l}\Delta b_{lp}}{D}.y_{pj} \\ \vdots \\ \dfrac{1}{D}y_{pj} \\ \vdots \\ y_{mj} - \dfrac{\beta_{ml}\Delta b_{lp}}{D}.y_{pj} \end{bmatrix}$$

$$= (c_{B1}, c_{B2}, \dots, c_{Bp}, \dots, c_{Bm}). \begin{bmatrix} y_{1j} - \dfrac{\beta_{1l}\Delta b_{lp}}{D}.y_{pj} \\ y_{2j} - \dfrac{\beta_{2l}\Delta b_{pj}}{D} \\ \vdots \\ \dfrac{1}{D}y_{pj} \\ y_{mj} - \dfrac{\beta_{ml}\Delta b_{lp}}{D}.y_{pj} \end{bmatrix} \xleftarrow[\text{p-th row}]{lp.y_{pj}}$$

$$= \sum_{i=1}^{m} c_{Bi}\left(y_{ij} - \frac{\beta_{il}\Delta b_{lp}}{D} y_{pj}\right) + \frac{c_{By}.y_{pj}}{D}$$

$$= \sum_{i=1}^{m}\left(c_{Bi}y_{ij} - \frac{c_{Bi}\beta_{il}\Delta b_{lp}}{D}y_{pj}\right) - c_{Bp}\left(y_{pj} - \frac{\beta_{pl}\Delta b_{lp}}{D}y_{pj}\right) + \frac{c_{Bp}.y_{pj}}{D}$$

$$= \sum_{i=1}^{m} c_{Bi}y_{ij} - \frac{1}{D}\sum_{i=1}^{m} c_{Bi}\beta_{il}\Delta b_{lp}y_{pj} - \frac{1}{D}[c_{Bp}.\ (D - b_{pl}\ \Delta\ b_{lp})\ y_{pj} - c_{Bp}\ y_{pj}]$$

$$= Z_j - \frac{1}{D}\sum_{i=1}^{m} c_{Bi}\beta_{il}\Delta b_{lp}y_{pj}$$

Since $\sum_{i=1}^{m} c_{Bi}y_{ij} = Z_j$ and $D = 1 + \beta_{pl}\ \Delta b_{lp}$

∴ For optimality of the solution $x_B{}^*$

$$\Delta_j^* = c_j - Z_j^* = c_j - Z_j + \frac{1}{D}\sum_{i=1}^{m} c_{Bi} \cdot \beta_{il} \Delta b_{lp} y_{pj} \le 0 \ \forall \ j \text{ not in the basis}$$

$$\Rightarrow D\,(c_j - Z_j) + \sum_{i=1}^{m} c_{Bi}\beta_{il}\Delta b_{lp} y_{pj} \le 0 \text{ assuming that } D = 1 + b_{pl}\,\Delta\, b_{lp} > 0$$

$$\Rightarrow (1 + b_{pl}\,\Delta b_{lp})\,(c_j - Z_j) + \sum_{i=1}^{m} c_{Bi}\beta_{il}\Delta b_{lp} y_{pj} \le 0$$

$$\Rightarrow \left[\beta_{pl}(c_j - Z_j) + \sum_{i=1}^{m} c_{Bi}\beta_{il} y_{pj}\right] \Delta\, b_{lp} \le -\,(c_j - Z_j)$$

$$\Rightarrow Q_j\ \Delta\ b_{lp} \le -\,\Delta_j$$

$$\text{where } Q_j = b_{pl}\,(c_j - Z_j) + \sum_{i=1}^{m} c_{Bi}\beta_{il} y_{pj}$$

$$= b_{pl}\,\Delta_j + \left[\sum_{i=1}^{m} c_{Bi}\beta_{il}\right] y_{pl}$$

$$\therefore \qquad \Delta\, b_{lp} \le \frac{-\Delta_j}{Q_j}, \text{ for } Q_j > 0$$

$$\text{and} \qquad \Delta\, b_{lp} \ge \frac{-\Delta_j}{Q_j}, \text{ for } Q_j < 0.$$

Hence the range of $\Delta\, a_{lk}$ $(= \Delta\, b_{lp})$ *so that the solution remains optimal is given by*

$$\text{Max.}\left[-\frac{\Delta_j}{Q_j < 0}\right] \le \Delta\, a_{lk} \le \text{Min}\left[\frac{-\Delta_j}{Q_j > 0}\right] \qquad \text{...(8)}$$

for all j not in the basis.

If no $Q_j < 0$, there is no lower bound to $\Delta\, a_{lk}$, and if no $Q_j > 0$, there is no upper bound to $\Delta\, a_{lk}$.

Hence a change Δa_{lk} in a_{lk} (an element of basis matrix B) for the solution to remain feasible and optimal can be made such that (6), (7) and (8) are satisfied.

To Find Limits of Variation of a_{1k} for the Feasibility of the Solution

From (8) of 4, the range of variation of $a_{lk} = a_{11}$ for the feasibility of the solution is given by

$$\text{Max.}\left[\frac{x_{Bi}}{P_i < 0}\right] \le \Delta\, a_{lk} \le \text{Min.}\left[\frac{x_{Bi}}{P_i > 0}\right] \qquad \text{...(1)}$$

where $D = 1 + \beta_{pl}\ \Delta\ \beta_{lp} \geq 0$, $P_i = \beta_{il}\ x_{Bp} - \beta_{pl}\ x_{Bi}$.

Here $a_{lk} = a_{11} \in B$. Since $a_{11} \in \alpha_1\ (= Y_1) = y_2$

$\therefore a_{lk} = a_{11} = b_{12} = b_{lp}$ *i.e.*, $l = 1$ and $k = 1$, $p = 2$

and $\quad i = 1, 2, 3$. (Since there are only three rows in A)

$x_{B1} = 56/27$, $x_{B2} = 5/3$, $x_{B3} = 5/27$.

$\therefore P_1 = \beta_{11}\ x_{B2} - \beta_{21}\ x_{B1} = (5/9).\ (5/3) - 0 = 25/27 > 0$

$P_2 = \beta_{21}\ x_{B2} - \beta_{21}\ x_{B2} = 0.5/3 - 0 = 0$

$P_3 = \beta_{31}\ x_{B2} - \beta_{21}\ x_{B3} = -\ 1/9.\ 5/3 - 0 = -\ 5/27 < 0.$

$\therefore$ from (1), we have

$$\text{Max.}\left[\frac{x_{B3}}{P_3 < 0}\right] \leq \Delta\ a_{11} \leq \text{Min.}\left[\frac{x_{B1}}{P_1 > 0}\right]$$

$$\Rightarrow \qquad \frac{5/27}{5/27} \leq \Delta a_{11} \leq \frac{56/27}{25/27}$$

$$\Rightarrow \qquad -1 \leq \Delta\ a_{11} \leq 56/25 \qquad \text{...(A)}$$

Also $D = 1 + \beta_{pl}\ \Delta\ a_{lk} = 1 \geq 0$, holds as $\beta_{pl} = \beta_{21} = 0$.

Again from (10) 4, the range of variation of $a_{lk} = a_{11}$ for the optimality of the solution is given by

$$\text{Max.}\left[\frac{-\Delta_j}{Q_j < 0}\right] \leq \Delta a_{lk} \leq \text{Min.}\left[\frac{-\Delta_j}{Q_j > 0}\right] \qquad \text{...(2)}$$

where $Q_j = \beta_{pl}\ \Delta_j + \left(\sum_{i=1}^{m} c_{Bi}\beta_{il}\right) y_{pj}$ for all j not in the basis

Here $Q_j = \beta_{21}\ \Delta_j + \left(\sum_{i=1}^{3} c_{Bi}\beta_{il}\right) y_{2j}$

$= \beta_{21}\ \Delta_j + (c_{B1}\ \beta_{11} + c_{B2}\ \beta_{21} + c_{B3}\ \beta_{31})\ y_{2j}$

$= 0 + (3.5/9 + 10.0 - 5.1/9)\ y_{2j} = (10/9)\ y_{2j}$ for all $j = 3, 5, 6, 7$ not in the basis.

$\therefore Q_3 = 10/9.y_{23} = 10/9.2/3 = 20/27 > 0$

$Q_5 = 10/9.y_{25} = 10/9.0 = 0$

$Q_6 = 10/9.y_{26} = 10/9.1/3 = 10/27 > 0$

$Q_7 = 10/9.y_{27} = 10/9.0 = 0.$

Since no $Q_j < 0$

$\therefore$ There is no lower bound to $a_{lk} = (a_{11})$

$\therefore$ from (2), we have

$$-\infty < \Delta a_{11} \le \text{Min.}\left\{-\frac{\Delta_3}{Q_3}, -\frac{\Delta_6}{Q_6}\right\}$$

$$\Rightarrow \quad -\infty < \Delta a_{11} \le \text{Min.}\left\{\frac{82/27}{20/27}, \frac{80/27}{10/27}\right\}$$

$$\Rightarrow \quad -\infty < \Delta a_{11} \le 41/10. \qquad \text{...(B)}$$

Since the range for Δa_{11} so that the solution remains optimal and feasible is given (A) and (B) both

∴ Both (A) and (B) are satisfied

if $\quad -1 \le \Delta a_{11} \le 56/25.$

Since $\quad a_{11} = 1$

∴ limits of variation of a_{11} are

$$-1 + 1 \le a_{11} \le 56/25 + 1$$

$$\Rightarrow \quad 0 \le a_{11} \le 81/25.$$

To find limits of variation of a_{23}. Here $a_{23} \notin \alpha_3\ (Y_3)$ which is not in B. From (2), 4 the range of Δa_{lk} (change in $a_{lk} \notin$ B) so that the solution remains optimal and feasible is given by

$$\left[\frac{\Delta_k}{\left(\sum_{i=1}^{m} c_{Bi}\beta_{il}\right) > 0}\right] \le \Delta a_{lk} \le \left[\frac{\Delta_k}{\left(\sum_{i=1}^{m} c_{Bi}\beta_{il}\right)} < 0\right] \qquad \text{...(3)}$$

Here $a_{lk} = a_{23} = 2$

∴ $l = 2,\ k = 3.$

$\Delta_k = \Delta_3 = -\,82/27,\ m = 3$ (number of constraints)

$$\sum_{i=1}^{m} c_{Bi}\beta_{il} = \sum_{i=1}^{3} c_{Bi}\beta_{i2} = c_{B1}\,\beta_{12} + c_{B2}\,\beta_{22} + c_{B3}\,\beta_{32}$$

$$= 3.\,(-\,5/27) + 10.\,(1/3) + 5\,(1/27) = 80/27 > 0$$

∴ There is no upper bound to $\Delta\, a_{23}$.

∴ from (3), we have

$$\frac{\Delta_3}{\sum_{i=1}^{3} c_{Bi}\beta_{i2} > 0} \le \Delta a_{23} < \infty$$

$$\Rightarrow \quad \frac{-82/27}{80/27} \le \Delta a_{23} < \infty$$

$$\Rightarrow \quad -\,41/40 \le \Delta\, a_{23} < \infty.$$

Since $a_{23} = 2$,

$\therefore$ limits of variation of a_{23} are

$$-41/40 + 2 \leq a_{23} < \infty + 2 \Rightarrow 39/40 \leq a_{23} < \infty.$$

ANALYSIS PROBLEM

Analysis problem is the study of sensitivity of the optimal solution of an LP problem due to discrete variations (changes) in its parameters. The degree of sensitivity of the solution due to these variations can range from no change at all to a substantial change in the optimal solution of the given LP problem. Thus, in sensitivity analysis, we determine the range over which the LP model parameters can change without affecting the current optimal solution. For this, instead of resolving the entire problem as a new problem with new parameters, we may consider the original optimal solution as an initial solution for the purpose of knowing the ranges, both lower and upper, within which a parameter may assume a value.

Different categories of parameter changes in the original LP model, discussed in this chapter include:

(i) Consumption of resources per unit of decision variables x_j (coefficients of decision variables on the left-hand side of constraints, a_{ij}).

(ii) Addition of a new variable to the existing list of variables in LP problem.

(iii) Addition of a new constraint to the original LP problem constraints.

(iv) Profit (or cost) per unit (c_j) associated with both basic and non-basic decision variables (coefficients in the objective function).

(v) Availability of resources (right-hand side of constants, b_i).

Change in the Availability of Resources (b_1)

Case I : *When a slack variation is in the basis (column B)*

When a slack variable is present in solution mix column B of optimal simplex table, the procedure for finding the range of variation for the corresponding right hand of the constraint is as follows:

Lower limit = Original value ù Solution value of slack variable

Upper limit = Infinity (∞)

Case II : *When slack variation is not in the solution mix*

The optimal simplex table, $c_j - z_j$ numbers (ignoring negative sign) corresponding to the slack variable columns represent shadow prices of the available resources. Shadow pricing provides important information in terms of change in the value of objective function from an increase of one unit of a scarce resource. Another important information which can be obtained from shadow pricing is in terms of resource value that should be increased in order to realize the best marginal increase in the value of the objective function.

Optimal simplex Table 3.1 is reproduced as Table 2.1. Recall that slack variable s_1 represents slack availability of manpower resource and s_2 the Unused raw material. We cannot add an unlimited number of units of resource without violating one of the problem constraints. For example, once we understand and compute the shadow price for an additional hour of manpower time ($c_4 - z_4$ = Rs. 10/3), we need to determine how many hours we can actually add or remove from manpower so as to increase profit and still have the same shadow price. This process in linear programming involves determining the range over which shadow prices will remain valid.

In Table 3.1, there is no slack variable in the solution mix column B. The procedure for finding the range for 'resource values' within which current optimal solution remains unchanged is summarized below.

1. Treat the slack variable corresponding to resource value as if it was an entering variable in the solution mix. For this, calculate exchange ratio (minimum ratio) for every row.

$$\text{Exchange ratio} = \frac{\text{Solution value, } x_B}{\begin{array}{c}\text{Exchange(input–Output) coefficients}\\ \text{in a slack variable column}\end{array}}$$

2. Find both lower and upper sensitivity limits.

Lower limit = Original value – Least (smallest) positive ratio or $-\infty$ (if no ratio is positive)

Upper limit = Original value + Smallest absolute negative ratio or ∞ (if no ratio is negative)

Table 2.1

		$c_j \rightarrow$	*4*	*6*	*2*	*0*	*0*
C_B	*Variations in Basis*	*Solution Values*	x_1	x_2	x_3	s_1	s_2
		B	b (= x_B)				
4	x_1	1	1	0	–1	4/3	–1/3
6	x_2	2	0	1	2	–1/3	1/3
Z = 16		z_j	4	6	8	10/3	2/3
		$c_j - z_j$	0	0	–6	–10/3	–2/3

To illustrate the method of finding the range of variation in the availability of resources, we repeat s_1 column and the *Solution values* column from Table 3.1 and calculate ratios as shown below:

Variations in Basis B	*Solution Values b (= x_B)*	*Exchange Coefficients in s_1-column*	*Exchange Ratio*	*Exchanges Coefficients in s_2-column*	*Exchange Ratio*
x_1	1	4/3	1/(4/3)=3/4	–1/3	1/(–1/3)=–3
x_2	2	–1/3	2/(–1/3)=–6	1/3	2/(1/3)=6

The smallest positive ratio (3/4) indicates as to by how many hours can manpower time resource be decreased (reduced) while smallest absolute negative ratio indicates as to by how many hours can this resource be increased (added) without changing the current optimal product mix. Thus, the shadow price for manpower resource (Rs. 10/3) and raw material resource (Rs. 2/3) are valid over the range as given below:

Solution Mix	Lower Limit	Upper Limit
Manpower (x_1)	3 – (3/4) = 9/4	3 + 6 = 9
Raw material (x_2)	9 – 6 = 3	9 + 6 = 15.

***Case III** : Changes in right-hand side when constraints are mixed type*

(i) When surplus variable is not in the basis (column B)

Lower limits = Original value – Smallest absolute value of negative exchange ratios or $-\infty$ (if no ratio is negative)

Upper limit = Original value + Smallest positive minimum ratio or (if no ratio is positive)

(ii) When surplus variable is in the basis (column B)

Lower limit = Minus infinity ($-\infty$)

Upper limit = Original value + Solution value of surplus variable.

Alternative Methods

1. Since b_1 values are not associated with the calculation of $c_j - z_j$ values, therefore, any change in the right hand-side of the constraints does not affect the optimality condition [$c_j - z_j = c_j - c_B B^{-1} a_j \leq 0$ (maximization case)]. However, it affects the values of basic variables and the value of the objective function Z (= $c_B x_B$) because in the determination of solution values, ($x_B = B^{-1} b$), value of resource (i.e. b) is involved. Thus, if resource b_k is changed to $b_k + \Delta b_k$, then the new values of resources becomes

$$\begin{aligned} b^* &= (b_1, b_2, ..., b_k + \Delta b_k, ..., b_m) \\ &= (b_1, b_2, ..., b_m) + (0, 0, ..., \Delta b_k, ..., 0) \\ &= b + \Delta b_k \end{aligned}$$

The range of values within which Δb_k can vary without affecting the optimality of the current solution is determined as follows:

$$\begin{aligned} x^*_B &= B^{-1}(b + \Delta b) = B^{-1} + B^{-1}(\Delta b) \\ &= B^{-1}b + B^{-1}(0, 0, ... \Delta b_k, ..., 0) \\ &= x_B + (\beta_1, \beta_2, ..., \beta_k)(0,0, ... \Delta b_k, ..., 0) \\ &= x_B + \beta_k(\Delta b_k) \end{aligned}$$

$$x^*_{Bi} = x_{Bi} + \beta_{ik}(\Delta b_k);\ \text{for ith basic variable}$$

where β_{ik} is the (i, k) element of B^{-1} and is the kth column vector of B^{-1}. In order to maintain the feasibility of the solution at each iteration, the solution values, x_B must be non-negative. That is, we must have

$$x^*_B = x_{Bi} + \beta_{ik}(\Delta b_k) \leq 0;\ i = 1, 2, ..., m$$

Hence, the range of variation in b_k can be obtained by solving the following system of inequalities

$$\min\left\{\frac{x_{Bi}}{\beta_{ik} < 0}\right\} \geq \Delta b_k \ \text{Max}\left\{\frac{-x_{Bi}}{\beta_{ik} > 0}\right\}$$

2. The range of variation in the availability of resources (b_i), can also be obtained by using condition of feasibility of the current optimal solution, i.e. $x_B = B^{-1}b \geq 0$,

where B^{-1} = matrix of coefficients corresponding to slack variables in the optimal simplex table

βb_k = amount of change in the resource k

x_B = basic variables appearing in B-column of simplex table.

Remarks:

1. If one or more entries in the x_B-column of the simplex table are negative, the dual simplex method can be used to get an optimal solution to the new problem by maintaining feasibility.
2. A resource whose shadow price is bigger in comparison to others, should be increased first to ensure the best marginal increase in the objective function value.

Change in Objective Function Coefficient (c_j)

Suppose the coefficient c, in the objective function of an LP model represents either profit or cost per unit of an activity (variable) x_j. Then the question that may arise is: *What happens to the optimal solution and the objective function value when this coefficient is changed*? For example, let Rs. 10 be the per unit profit coefficient of a particular variable in the objective function. After obtaining an optimal solution to the LP problem with Rs. 10 as one of the objective function coefficients, the decision-maker realises that its true value might be any value between Rs. 9 and Rs. 11 per unit. The test of sensitivity of the objective function value with respect to this coefficient (or on any other such coefficients) determines the range (both lower and upper) of values within which each c_j $(j = 1, 2, ..., n)$ can lie without changing the current optimal solution. Such an analysis can help the decision-maker in deciding whether resources from other activities (variables) should be diverted to (diverted away from) a more profitable (or less profitable) activity.

Changes in the profit or cost coefficients (contributions) in the objective function can occur for a *basic variable* (the variable in the solution mix) or a non-basic variable (the variable not in the solution mix). The sensitivity ranges for these variables are determined differently. Thus, these two cases will be discussed separately.

Given an optimal basic feasible solution $[x_B = B^{-1}b]$ with basis matrix B, suppose that the coefficient c_k of a variable x_k is changed from c_k to $c_k + \Delta c_k$. Since the optimal basic feasible solution or solution values appeared in the 'x_B' column of the simplex table do not involve cost (or profit) coefficients, c_j in their calculations, it will remain feasible even for any change in the coefficients in the objective function. However, such a change in coefficients (c_j's) may affect the optimality of this solution. In other words, as $c_k - z_k$ $(= c_k - c_B B^{-1} a_k)$ involves coefficients c_k, the effect of this change will be seen in the $c_j - z_j$ row of the optimal simplex table.

Case I : *Change in the coefficient of a basic variation*

In the maximization LP problem the change in the coefficient, say c_k, of a basic variable x_k affects the $c_j - z_j$ values corresponding to all non-basic variables in the simplex table. It is because the coefficient c_k is listed in the c_B column of the simplex table and affects the calculation of z_j values.

The sensitivity limits for the contribution per unit of a basic variable are calculated as under:

Lower limit = Original value, c_k – {Lowest absolute value of improvement ratio or $-\infty$ (if no ratio is negative)}

Upper limit = Original value, c_k + {Lowest positive value of improvement ratio or ∞ (if no ratio is positive)} where Improvement ratio $= \dfrac{\text{Per unit improvement value}}{\text{input - output coefficient in the variable row}}$

$$= \frac{c_j - z_j}{a_{kj}}$$

Remark: While performing sensitivity analysis, the *artificial variable columns in the simplex table are ignored.* Any exchange of quantities between basic variables and an artificial variable make no sense, because an artificial variable has no economic interpretation. Thus, improvement ratios using coefficients corresponding to artificial variables should not be considered.

Case II : *Change in coefficient of a non-basic variation in cost minimization problem*

The procedure for calculating sensitivity limits to a cost minimization LP problem, where the objective function coefficients are unit costs is identical to the Case I discussed above. In this case, the unit cost coefficient can be increased to any arbitrary level but it cannot be decreased by more than per unit inprovement value without making it eligible so that a non-basic variable can be entered into the new solution mix. The sensitivity limits can be calculated as:

Lower limit = Original value – Unit improvement value

Upper limit = Infinity (∞)

Alternative Method

If we add Δc_k to the objective function coefficient c_k of the basic variable x_k, then its coefficient c_{Bk} will become $c^*_{Bk} = c_{Bk} + \Delta c_{Bk}$. Now $c_j - z_j$ is calculated as follows:

$$c_j - z^*_j = c_j - \sum_{i=1}^{m} c_{Bi} y_{ij} = c_j - \left\{ \sum_{i \neq k}^{m} c_{Bi} y_{ij} + \left(c^*_{Bk} + \Delta c_{Bk} \right) y_{kj} \right\}$$

$$= c_j - \left\{ \sum_{i \neq k}^{m} c_{Bi} y_{ij} + c_{Bk}\, y_{kj} \right\}$$

$$= c_j - \{z_j + \Delta c_{Bk} y_{kj}\} = (c_j - z_j) - \Delta c_{Bk} y_{kj}$$

For a current basic feasible solution to remain optimal for a maximization LP problem, we must have $c_j - z^*_j \leq 0$. That is,

$$c_j - z^*_j = (c_j - z_j) - \Delta c_{Bk} y_{kj} \leq 0$$

or $\quad c_j - z_j \, \Delta \leq c_{Bk} y_{kj}$

or $$\begin{cases}(c_j - z_j)/y_{kj} \le \Delta c_{Bk} \text{ when } y_{kj} > 0 \\ (c_j - z_j)/y_{kj} \ge \Delta c_{Bk} \text{ when } y_{kj} > 0\end{cases}$$

Hence, the value of Δc_{Bk} which satisfies the optimality criterion can be determined by solving the following system of linear inequalities:

$$\text{Min}\left\{\frac{c_j - z_j}{y_{kj} < 0}\right\} \ge \Delta c_{Bk} \ge \text{Max}\left\{\frac{c_j - z_j}{y_{kj} > 0}\right\}$$

Here it may be noted that y_{kj},'s are entries in the non-basic variable columns (i.e. variables) of the optimal simplex table.

The new value of the objective function becomes

$$Z^* = \sum_{i \neq k}^{m} c_{Bi}x_{Bi} + (c_{Bk} + \Delta c_{Bk})x_{Bk} = \sum_{i \neq k}^{m} c_{Bi}x_{Bi} + \Delta c_{Bk}x_{Bk} = Z + \Delta c_{Bk}x_{Bk}$$

Hence if Δc_{Bk} satisfies inequality (1), then the optimal solution will remain unchanged but the value of Z will be improve by an amount $\Delta c_{Bk}x_{Bk}$..

***Case III :** Change in the coefficient of a non-basic variation*

The current optimal solution for a maximization LP problem will remain optimal as long as all $c_j - z_j \le 0$, for all j. Let c_k be the coefficient of a non-basic variable x_k in the objective function. Since c_k is the coefficient of non-basic variable x_k, therefore, it does not affect any of the c_j values listed in the 'c_B' column of simplex table, associated with basic variables. Since the calculation of $z_j = c_B B^{-1} a_j$ values do not, involve c_j, therefore changes in c_j does not alter z_j values and hence $(c_j - z_j)$ values remain unchanged except $c_k - z_k$ value due to change in c_k. In other words, any change in this coefficient does not affect feasibility of the optimal solution. This means that unit profit of x_k can be lowered to any level without causing the optimal solution to change. But any increase in its unit profit beyond a certain level (i.e. upper limit) should make this variable eligible to be a basic variable in the new solution mix. Obviously, then $c_k - z_k$ will no longer be negative.

To retain optimality of the current optimal solution for a change Δc_k in c_k, we must have $(c_k + \Delta c_k) - z_k \le 0$ or $c_k + \Delta c_k \le z_k$. Hence, for an LP problem with an objective function of maximization type, the value of c_k may be increased up to the value of z_k, and decrease to negative infinity $(-\infty)$ without affecting the optimal solution.

Changes in the Input-Out Coefficients (a_{ij}'s)

Suppose that the elements of coefficient matrix A are changed. Then two cases arise

(i) Change in a coefficient, when variable is a basic variable, and

(ii) Change in a coefficient, when variable is a non-basic variable.

Case I : When a non-basic column $a_k \notin B$ changed to a^*_k, the only effect of such change will be on the optimality condition. Thus the solution will remain optimal, if

$$c_k - z^*_k = c_k - c_B B^{-1} a^*_k \le 0$$

otherwise the simplex method is continued, after column k of the simplex table is updated, by introducing the non-base variable x_k into the basis.

However, the range for the discrete change Δa_{ij} in the coefficient of non-basic variable x_j in the constraint, i can be determined by solving following linear inequalities:

$$\min \left\{ \frac{c_j - z_j}{c_B \beta_i > 0} \right\} \le \Delta a_{rj} \le \text{Max} \left\{ \frac{c_j - z_j}{c_B \beta_i < 0} \right\}$$

Here β_i is the ith column B^{-1}.

If $c_B\ \beta_i = 0$,

then Δa_{ij}. is unrestricted in sign.

Alternative Method

The change in the coefficients (a_{ij}'s) values associated with non-basic variables in the optimal simplex table can be analysed by forming a corresponding dual constraint from the original set of constraints:

$$\sum_{i=1}^{m} a_{ij}\, y_i \ge c_j; \text{ for } x_j \text{ non-basic variable.}$$

The values of dual variable y_i's can be obtained from the optimal simplex table. The reason behind this dual constraint formulation is that an activity is considered as fully undertaken provided all the marginal values (shadow prices) of its resources become equal to its per unit contribution to total profit.

Case II : Suppose a basic variable column $a_k \in B$ is changed to a^*_k Then conditions to maintain both feasibility and optimality of the current optimal solution are:

(a) $$\underset{k \ne p}{\text{Max}} \left\{ \frac{-x_{Bk}}{x_{Bk}\beta_{pi} - x_{Bp}\,\beta_{ki} > 0} \right\} \le \Delta a_{ij} \le \underset{k \ne p}{\text{Min}} \left\{ \frac{-x_{Bk}}{x_{Bk}\beta_{pi} - x_{Bp}\,\beta_{ki} < 0} \right\}$$

(b) $$\text{Max} \left\{ \frac{c_j - z_j}{(c_j - z_j)\beta_{pi} - y_{pj}\, c_B\, \beta_i > 0} \right\} \le \Delta a_{ij} \le \text{Min} \left\{ \frac{c_j - z_j}{(c_j - z_j)\beta_{pi} - y_{pj}\, c_B\, \beta_i < 0} \right\}$$

ANALYSIS PROBLEM IN ASSIGNMENT PROBLEMS

There is very little scope for sensitivity analysis in assignment problem because of its structure. Modest alterations in the conditions (such as one being able to do two jobs) can be considered by repeating the man's row and adding a dummy a column to square up the matrix. The problem of not assigning a particular job (ith) to particular facility (jth) can be solved by taking a very large cost (∞) of this assignment.

Also, the addition of a constant throughout any row or column does not affect the optimal solution of the assignment problem.

TRAVELLING PROBLEM OF SALESMAN

Suppose a salesman wants to visit a certain number of cities.

The distances (or time or cost) of journey between every pair of cities allotted to him. His problem is to choose such a rout which starts from his home city, passes through each city once and only once and returns this home city in the shortest possible distance (or in least time or at least cost).

The above problem may be classified in two forms:

(i) Symmetrical

The distance (or time or cost) between every pair of cities is independent of the direction of journey the problem is said to be symmetrical.

(ii) Asymmetrical

One or more pairs of cities, the distance (or time or cost) changes with the direction, the problem is said to be asymmetrical. For example, it takes longer time while going up-hill from city A to B instead of coming down hill from city B to A. Similarly flying from East to West usually takes longer time than from West to East on account of prevailing winds.

Further we note that for two cities there is only one possible route *i.e.*, there is no choice.

In case of three cities, say A, B and C, one of them (say A) is the home base, there are two possible routes: A $\to$ B $\to$ C and A $\to$ C $\to$ B. For four cities there are 3! = 6 possible routes. In general, there are $(n - 1)!$ possible routes if there are n cities. Thus, practically it is impossible to find the best route by trying each one. That is why the travelling salesman problem is considered as a puzzle by the mathematicians. The best procedure to solve the problem is as if it were an assignment problem. We formulate the problem of travelling salesman in the form of an assignment problem with the additional restriction on his choice of route.

Formulation of a Travelling Salesman Problem as Assignment Problem

Let $x_{ij} = 1$, if the salesman goes directly from city A_j, to c_{ij} to A_j and zero otherwise. Also, let c_{ij} be the distance (or time or cost) from city A_i to city A_j. Then our problem is to minimize $Z = \sum\sum c_{ij}\, x_{ij}$ with one additional restriction that the x_{ij}'s must be so chosen that no city is visited i_j twice until the tour of all the cities is completed. In particular, he cannot go directly from city A_i to A_i itself. To avoid this possibility in the minimization process we adopt the convention $c_{ij} = \infty$ so that x_{ij} can never be unity when $i = j$. Also we note that only one $x_{ij} = 1$ for each value of i and j. The distance (or time or cost) matrix for this problem is given in the following table:

		To			
		A_1	A_2	...	A_n
	A_1	∞	c_{12}	...	c_{1n}
	A_2	c_{21}	∞	...	c_{2n}
	.	.	.	.	.
From	.	.	.	.	.
	.	.	.	.	.
	A_n	c_{n1}	c_{n2}	...	∞

We can omit the variable x_{ij} from the problem specification. Our problem is to determine a set of n elements of this matrix, one in each row and one in each column, so as to minimize the sum of the elements determined above.

Note: A problem similar to travelling salesman arises when n items say A_i, $i = 1, 2, \ldots n$, are to be produced on a machine in continuation, given that c_{ij} $(i, j = 1, 2, \ldots n)$ is the setup cost of the machine when item A_i is followed by A_j. Here two additional restrictions are imposed. One restriction is that we do not follow A_i again by A_i. The other restriction is that we do not produce an item again until all items are produced once.

Solution Procedure

The problem could be solved by assignment technique. In some cases (violation of additional restriction) we use the method of numeration by assignment the next minimum element of matrix in place of zero.

NEW CONSTRAINT PROBLEM ADDITION OF A

Let a new constraint be introduced to a L.P.P. whose optimal solution has been obtained. Here we assume that the additional constraint does not introduce any new variable with non-zero-price.

Let Z* be the optimal (maximal) value of the objective function for the new L.P.P. while it was Z for the original problem. Let Z* > Z. Since the new optimal solution satisfies the first m constraints as well as the additional new constraint, therefore, it is also an optimal solution of the original problem which is contradiction to the fact that we already had an optimal solution to the original L.P.P. Here $Z^* \leq Z$. (Max.)

Thus, we have the following two cases:

Case I : If the optimal solution of the original L.P.P. satisfies the new constraint, it is also an optimal solution of the new L.P.P. In this case the adulterant is redundant.

Case II : If the optimal solution of the original L.P.P. does not satisfy the new constraint, a new optimal solution of the new L.P. problem must be obtained as follows:

To Find the New Optimal Solution of the New Enlarged Problem

Let B and B_1 be the optimal basis of the original and the enlarged L.P. problem respectively. Clearly B_1 is the square matrix of order (m + 1) if B the square matrix of order m.

∴ We can write

$$B_1 = \begin{bmatrix} B & 0 \\ \alpha & \pm 1 \end{bmatrix} \qquad ...(1)$$

The last column of B_1 corresponds to the slack, surplus or artificial vector associated with the additional new constraints and α is the row vector of the coefficients, in the new constraint, of the variables which correspond to the vectors in the optimal basis B.

Since B^{-1} exist and is known therefore the inverse of B_1 is given by

$$B_1^{-1} = \begin{bmatrix} B^{-1} & 0 \\ \mp \alpha B^{-1} & \pm 1 \end{bmatrix} \qquad ...(2)$$

Let a_{m+1}, j be the coefficient of x_j in the new (m + 1)th constraint and α_j^* the column vector of the coefficients of x_j in the enlarged problem. Also if Y_j^* and Z_j^* are Y_j and Z_j for the new problem, then we have

$$Y_j^* = B_1^{-1}.\alpha_j^* = \begin{bmatrix} B^{-1} & 0 \\ \mp \alpha B^{-1} & \pm 1 \end{bmatrix} \begin{bmatrix} \alpha_j \\ a_{m+1.j} \end{bmatrix}$$

$$= \begin{bmatrix} Y_j \\ \mp \alpha B^{-1} \alpha_j \pm a_{m+1,j} \end{bmatrix}$$

$$\Rightarrow Y_j^* = \begin{bmatrix} Y_j \\ \mp \alpha Y_j \pm a_{m+1,j} \end{bmatrix}$$

Now $Z_j^* = c_{B1} \; Y_j^* = (c_B, c_{B.m+1}) \begin{bmatrix} Y_j \\ \mp \alpha Y_j \pm a_{m+1,j} \end{bmatrix}$

or $Z_j^* = c_{B1} \; Y_j + c_{B, \; m+1} \cdot (\pm \alpha Y_j \pm a_{m+1,j})$ ...(3)

(i) *If slack or surplus variable is introduced in the additional constraint.*

In this case $c_{B,m+1} = 0$

$\therefore$ from (3),

$$Z_j^* = c_B \; Y_j = Z_j$$

$\therefore c_j - Z_j^* = c_j - Z_j$

$\therefore c_j - Z_j^*$

remains unchanged is this case. Since the optimal solution of the original problem does not satisfy the new constraint, therefore the slack or surplus variable introduced in the new constraint is negative. *Hence we can apply the dual simplex algorithm to find an optimal feasible solution of the new problem.*

(ii) *If the artificial variable is introduced in the additional constraint i.e., if the additional constraint is a perfect equality.* In this case the additional vector is an artificial vector. Now there are two possibilities:

(a) *If the artificial variable in the basic solution is negative,* then assigning a price to the artificial variable we can use the dual simplex algorithm for the removal of the artificial variable from the basis.

(b) *If the artificial variable in the basis solution is positive,* then assigning a price –M to the artificial variable we can use the standard simplex method for the removal of the artificial variable from the basis. It is important to note that in this case $c_j - Z_j$ will be changed.

NEW VARIABLE PROBLEM ADDITION OF A

If a new variable is introduced in a L.P.P. whose optimal solution has been obtained, then the solution of the problem will remain feasible. Addition of an extra variable x_{n+1} to the problem will introduce an extra column say α_{n+1} to the coefficient matrix A and an extra cost c_{n+1} will be introduced in the price vector c. Thus the addition of this extra variable may affect the optimality of the problem.

For the same basis B the solution $\mathbf{x}_B$ of the original problem will remain optimal (maxima) if

$$c_{n+1} - Z_{n+1} \leq 0$$

$$\Rightarrow c_{n+1} - c_B\, B^{-1}\, \alpha_{n+1} \leq 0. \qquad ...(1)$$

i.e., if (1) is satisfied then the new variable becomes just like a non-basis variable having zero value.

If $c_{n+1} - Z_{n+1} > 0$, then the solution $\mathbf{x}_B$ is no more optimal for the new problem and can be improved by introducing α_{n+1} in the basis. Here we can start with the last simplex table giving the optimal feasible solution of the original problem by introducing one more column corresponding to the variable x_{n+1}.

SOLVED EXAMPLES

Example 1(a):

The linear programming problem is

$$Max\ Z = 3x_1 + 5x_2$$

$$s.t. \qquad x_1 + x_2 \leq 1$$

$$2x_2 + 3x_2 \leq 1$$

$$and \qquad x_1, x_2 \geq 0.$$

Obtain the variations in cj (j = 1, 2) which are permitted without changing the optimal solution.

Solution:

Proceeding as usual, the *final simplex table* giving the optimal solution of the given L.P.P. is as follows.

		c_j	3	5	0	0
B	c_B	x_B	Y_1	$Y_2\ (\beta_2)$	$Y_3\ (\beta_1)$	Y_4
Y_3	0	2/3	1/3	0	1	–1/3
Y_2	5	1/3	2/3	1	0	1/3
$Z = c_B x_B = 5/3$		Δ_j	–1/3	0	0	–5/3

From the above table the optimal solution, we have

$x_1 = 0,\ x_2 = 1/3$, Max. $Z = 5/3$.

Here $c_B = (c_{B1}, c_{B2}) = (0, 5) = (c_3, c_2)$.

To find variation in c_1. Here c_1 does not belong to c_B.

∴ From (1), 2, the change Δc_1 in c_1, so that the solution remains optimal, is given by

$$\Delta c_1 \le Z_1 - c_1 = -\Delta_1.$$

$$\Rightarrow \quad \Delta c_1 \le 1/3.$$

∴ The range over which c_1 can vary maintaining the optimality of the solution given in above table is

$$-\infty < c_1 \le c_1 + \Delta c_1$$

$$\Rightarrow \quad -\infty < c_1 \le 3 + 1/3$$

$$\Rightarrow \quad -\infty < c_1 \le 10/3.$$

The value of the objective function will not change in this case.

To find variation in c_2. Here $c_2 = c_{Bk} = c_{B2} = 5 \in c_B$

∴ From (3), 2 the range of Δc_{B2} (change in c_{B2}) is given by

$$\underset{y_{kj}>0}{\text{Max}}\left[\frac{c_j - Z_j}{y_{kj}}\right] \le \Delta c_{Bk} \le \underset{y_{kj}<0}{\text{Min}}\left[\frac{c_j - Z_j}{y_{kj}}\right].$$

Here $\quad k = 2$ and $y_{21} = 2/3$,

$$y_{24} = 1/3 > 0.$$

Here we cannot consider y_{22} and y_{23} as α_2 and α_3 are in the basis.

Since no $y_{2j} < 0$,

∴ There is no upper bound to Δc_{B2}.

The change in Δc_{B2} is given by

$$\text{Max.}\left[\frac{\Delta_1}{y_{21}}, \frac{\Delta_4}{y_{24}}\right] \le \Delta c_{b2} < \infty$$

$$\text{Max.}\left[\frac{-1/3}{2/3}, \frac{-5/3}{1/3}\right] \le \Delta c_{B2} < \infty$$

$$\Rightarrow \quad -1/2 \le \Delta c_{B2} < \infty$$

$$\Rightarrow \quad c_{B2} - 1/2 \le c_2 < c_{B2} + \infty$$

$$\Rightarrow \quad 5 - 1/2 \le c_2 < 5 + \infty$$

$$\Rightarrow \quad 9/2 \le c_2 < \infty.$$

Example 1(b):

A salesman estimate that the following would be the cost on his route visiting the six cities as shown.

		To city					
		1	*2*	*3*	*4*	*5*	*6*
	1	∞	*20*	*23*	*27*	*29*	*24*
	2	*21*	∞	*19*	*26*	*31*	*24*
From city	*3*	*26*	*28*	∞	*15*	*36*	*26*
	4	*25*	*16*	*25*	∞	*23*	*18*
	5	*23*	*20*	*23*	*31*	∞	*10*
	6	*27*	*18*	*12*	*35*	*16*	∞

The sales man can visit each of the cities once and only once. Determine the optimum sequence he should follow to minimize the total distance travelled. What as the total distance travelled?

Solution:

Following the usual procedure of assignment algorithms, we obtain Table 12.1 showing an optimum solution indicated by encircled zeros.

This table provides the optimum assignment schedule

$1 \to 3 \to 4 \to 2 \to 1$ and $5 \to 6 \to 5$

with distance according to this optimum route being 101.

Table

	1	2	3	4	5	6
1	∞	0	0	7	2	14
2	0	∞	0	10	8	8
3	6	13	∞	0	14	11
4	4	0	6	∞	0	2
5	8	30	10	21	∞	0
6	13	6	0	26	0	∞

This table does not provide the solution to the travelling salesman problem, as it gives $1 \to 3$, $3 \to 4$, $4 \to 2$, $2 \to 1$, while city 2 is not allowed to follow city 1 unless city 5 and city 6 processed.

Now we try to find the next best solution which also satisfies this extra restriction. The next minimum (non-zero) element in the matrix is 2. Therefore, we try to bring 2 into the solution. But the element 2 occurs at two places.

We shall consider both the cases separately until the acceptable solution is attained,

	1	2	3	4	5	6
1	∞	0	0	7	(2)	14
2	0	∞	0	10	8	8
3	6	13	∞	0	14	11
4	4	0	6	∞	0	2
5	8	30	10	21	∞	0
6	13	6	0	26	0	∞

We start with the element (1, 5) and make the assignment in this cell instead of zero assignment in the cell (1, 3). After making this assignment we see that no other assignment can be made in the first row and third column, and thus the resulting feasible solution will be

$1 \to 5,\ 5 \to 6,\ 6 \to 3,$

$3 \to 4,\ 4 \to 2,\ 2 \to 1.$

The selected elements for this solution are encircled in Table 12.2. The cost corresponding to this feasible solution is 2.

Again, if we make assignment in the cell (4, 6) instead of (4, 2) then no feasible solution is available in terms of zeros. Hence the best programme is

$1 \to 5 \to 6 \to 3 \to 4 \to 2 \to 1.$

The total set-up cost according to this route comes out to be 103.

Solve the travelling salesman problem given by the following data:

$c_{12} = 20,\ c_{13} = 4,$

$c_{14} = 10,$

$c_{23} = 5,\ c_{24} = 6,$

$c_{25} = 10,$

$c_{35} = 6$

and $c_{45} = 20,$

where $c_{ij} = c_{ij}$

and there is no route between cities i and j if a value for c_{ij} is not shown above.

Given the following matrix of set-up cost, show how to sequence production so as to minimize set-up cost per cycle:

		To				
		A	B	C	D	E
	A	∞	2	5	7	1
	B	6	∞	3	8	2
From	C	8	7	∞	4	7
	D	12	4	6	∞	5
	E	1	3	2	8	∞

A medical representative has to visit five stations A, B, C, D, and E. He does not want to visit any station twice before completing his four of all the stations, and wishes to return to the starting station. Costs of going from one stations to another are given below. Determine the optimal route:

	A	B	C	D	E
A	∞	2	4	7	1
B	5	∞	2	8	2
C	7	6	∞	4	6
D	10	3	5	∞	4
E	1	2	2	8	∞.

Example 1(c):

Given the following L.P.P.

$$\text{Max. } Z = -x_1 + 2x_2 - x_3$$

s.t.

$$3x_1 + x_2 - x_3 \le 10$$

$$-x_1 + 4x_2 + x_3 \ge 6$$

$$x_2 + x_3 \le 4$$

and $x_1, x_2, x_3 \ge 0.$

Find the separate range of b_1, b_2 and b_3 (the constants on the right hand sides of the constraints) consistent with the optimal solution.

Solution:

Introducing the slack variables x_4, x_6, surplus variables x_5 and artificial variables x_7, the given L.P.P. reduces to

$$\text{Max. } Z = -x_1 + 2x_2 - x_3 + 0,$$

$$_4 + 0.x_5 + 0.x_6 - Mx_7$$

s.t. $3x_1 + x_2 - x_3 + x_4 = 10$

$$-x_1 + 4x_2 + x_3 - x_5 + x_7 = 6$$
$$x_2 + x_3 + x_6 = 4$$
and $\quad x_1, \ldots, x_7 \geq 0.$

Proceeding as usual, the successive tables of simplex method are as follows

		c_j	−1	2	−1	0	0	−M		Min. ratio
B	c_B	x_B	Y_1	Y_2	Y_3	Y_4	Y_5	Y_6	A_1	x_B/Y_2
Y_4	0	10	3	1	−1	1	0	0	0	10/1
A_1	−M	6	−1	(4)	1	0	−1	0	1	6/4 ←
Y_6	0	4	0	1	1	0	0	1	0	4/1
$Z=c_B x_B=-6M$		Δ_j	−1 −M	2+4M ↑	−1+M	=	−M	0	0 ↓	x_B/Y_5
Y_4	0	17/2	13/4	0	−5/4	1	1/4	0		$\frac{17}{2}/\frac{1}{4}$
Y_2	2	3/2	−1/4	1	1/4	0	−1/4	0		$-\frac{5}{2}/\frac{1}{4}$ ←
Y_6	0	5/2	1/4	0	3/4	0	(1/4)	1		
$Z=c_B x_B=3$		Δ_j	−1/2	0	−3/2	0	1/2 ↑	0 ↓		
Y_4	0	6	3	0	−2	1	0	−1		
Y_2	2	4	0	1	1	0	0	−1		
Y_5	0	10	1	0	3	0	1	4		
$Z=c_B x_B=8$		Δ_j	−1	0	−3	0	0	−2		

We have Optimal solution is $x_1 = 0$, $x_2 = 4$ $x_3 = 0$, Max $Z = 8$.

Here $B = (\alpha_4, \alpha_2, \alpha_5)$, $b = (10, 6, 4)$

$$x_B = (x_4, x_2, x_5) = (x_{B1}, x_{B2}, x_{B3}) = (6, 4, 10),$$

$$\therefore B^{-1} = (\beta_1, \beta_2, \beta_3) = \begin{bmatrix} 1 & 0 & -1 \\ 0 & 1 & 1 \\ 0 & 0 & 4 \end{bmatrix}$$

To find variation in b_1. After changing b_1 to $b_1 + \Delta b_1$, the new requirement vector is given by

$$b_B^* = (10 + \Delta b_1, 6, 4)$$

From relation (2), 3 the range of Δb_1 consistent with the optimal solution is given by

$$\underset{\beta_{il} > 0}{\text{Max.}}\left[\frac{-x_{Bi}}{\beta_{il}}\right] \le \Delta b_1 \le \underset{\beta_{il} < 0}{\text{Min.}}\left[\frac{-x_{Bi}}{\beta_{il}}\right]$$

Here $b_{11} = 1 > 0$. Since no $b_{il} < 0$,

$\therefore$ there is no upper bound to Δb_1.

$\therefore$ We have Max. $\left[-\frac{x_{Bl}}{\beta_{11}}\right] \le \Delta b_1 < \infty$.

$$\Rightarrow \qquad -\frac{6}{1} \le \Delta b_1 < \infty.$$

$\therefore$ Range for b_1 is $-6 + 10 \le b_1 < 10 + \infty$

$$\Rightarrow \qquad 4 \le b_1 < \infty$$

To find variation in b_2.

After changing b_2 to $b_2 + \Delta b_2$, the new requirement vector is

$$b_B^{**} = (10, 6 + \Delta b_2, 4).$$

From relation (2), 3 the range of Δb_2 consistent with the optimal solution is given by

$$\underset{\beta_{i2} > 0}{\text{Max}}\left[\frac{-x_{Bi}}{\beta_{i2}}\right] \le \Delta b_2 \le \underset{\beta_{i2} < 0}{\text{Min.}}\left[\frac{-x_{Bi}}{\beta_{i2}}\right]$$

Here $\beta_{22} = 1 > 0$. Since no $\beta_{i2} < 0$. $\therefore$ there is no upper bound to Δb_2.

$\therefore$ We have Max. $\left[-\frac{x_{B2}}{\beta_{22}}\right] \le \Delta b_2 < \infty$

$$\Rightarrow \qquad \frac{-4}{1} \le \Delta b_2 < \infty$$

$\therefore$ Range of variation of b_2 is

$$-4 + 6 \le b_2 < \infty + 0$$

$$\Rightarrow \qquad 2 \le b_2 < \infty.$$

To find variation in b_3.

After changing b_3 to $b_3 + \Delta b_3$, the new requirement vector is $b_B = (10, 6, 4 + \Delta_3)$, from relation (2), 3 the range of Δb_3 consistent with the optimal solution is given by

$$\underset{\beta_{i3}>0}{\text{Max.}}\left[\frac{-x_{Bi}}{\beta_{i3}}\right]\leq \Delta b_3 \leq \underset{\beta_{i3}<0}{\text{Min.}}\left[\frac{-x_{Bi}}{\beta_{i3}}\right]$$

Here $\beta_{23} = 1$, $\beta_{33} = 4 > 0$ and $\beta_{13} = -1 < 0$.

$$\therefore \text{Max.}\left[\frac{-x_{B2}}{\beta_{23}}, -\frac{x_{B3}}{\beta_{33}}\right]\leq \Delta b_3 \leq \text{Min.}\left[-\frac{x_{B1}}{\beta_{13}}\right]$$

$$\Rightarrow \text{Max.}\left\{-\frac{4}{1}, -\frac{10}{4}\right\}\leq \Delta b_3 \leq \text{Min.}\left\{\frac{-6}{-1}\right\}$$

$$\Rightarrow -\frac{5}{2}\leq \Delta b_3 \leq 6$$

$\therefore$ Range of variation of b_3 is

$$4-\frac{5}{2}\leq b_3 \leq 4+6$$

$$\Rightarrow \frac{3}{2}\leq b_3 \leq 10.$$

Example 2(a):

Consider the following table which presents an optimal solution to some linear programming problem.

Table

B	c_B	c_j	2	4	1	3	2	0	0	0
		x_B	Y_1	Y_2	Y_3	Y_4	Y_5	Y_6	Y_7	Y_8
Y_1	2	3	1	0	0	–1	0	0.5	0.2	–1
Y_2	4	1	0	1	0	2	1	–1	0	0.5
Y_3	1	7	0	0	1	–1	–2	5	–0.3	2
$Z = c_B x_B = 17$		Δ_j	0	0	0	–2	0	–2	–0.1	–2

If the additional constraint

$$2x_1 + 3x_2 - x_3 + 2x_4 - 4x_5 \leq 5$$

were annexed to the system, would there be any change in the optimal solution? Justify your answer.

Solution:

From the table the optimal solution of the given L.P. problem is

$x_1 = 3$, $x_2 = 1$, $x_3 = 7$,

$$x_4 = 0 = x_5 = x_6 = x_7 = x_8$$

which also satisfies the new additional constraint

$$2x_1 + 3x_2 - x_3 + 2x_4 - 4x_5 \leq 5.$$

Thus, the optimal solution of the given problem will not be changed if we introduce the above constraint to it.

Hence, the additional constraint is redundant and the optimal sol. of the given L.P.P. is also the optimal solution of the new L.P.P.

Example 2(b):

Solve the travelling salesman problem given by the following data:

$c_{12} = 20,\ c_{13} = 4,\ c_{14} = 10,\ c_{23} = 5,\ c_{34} = 6$

$c_{25} = 10,\ c_{35} = 6,\ c_{45} = 20,$ *where* $c_{ij} = c_{ji}$,

and there is no route between cities i and j if the value for c_{ij} *is not given above.*

Solution:

Consider the given problem as an assignment problem.

Taking $c_{ij} = \infty$ for $i = j$, the cost matrix is as follows:

	1	2	3	4	5
1	∞	20	4	10	∞
2	20	∞	5	∞	10
3	4	5	∞	6	6
4	10	∞	6	∞	20
5	∞	10	6	20	∞

If there is no route between cities i and j then we have taken $c_{ij} = \infty$ to avoid the possibility of going from ith station to jth station.

We shall solve the problem by usual assignment algorithm. The following tables show the necessary steps for reaching the solution:

∞	15	(0)	4	∞	∞	12	(0)	1	∞	∞	12	(0)	⊗	∞
15	∞	⊗	∞	3	12	∞	⊗	∞	(0)	11	∞	⊗	∞	(0)
(0)	⊗	∞	⊗	⊗	(0)	⊗	∞	⊗	⊗	⊗	1	∞	(0)	1
4	∞	⊗	∞	12	1	∞	⊗	∞	9	(0)	∞	⊗	∞	9
∞	3	⊗	12	∞	∞	(0)	⊗	9	∞	∞	(0)	⊗	8	∞

Hence, the optimum solution of the assignment problem is:

$1 \rightarrow 3$, $3 \rightarrow 4$, $4 \rightarrow 1$, $2 \rightarrow 5$, $5 \rightarrow 2$.

But this is not the solution to the travelling salesman problem, as it is not allowed to go from city 4 to 1 without visiting the cities 2 and 5.

We try to find the 'next best' solution which satisfies the additional restriction. The smallest element other than zero is 1. So we try to bring 1 into the solution. Since the element 1 occurs at two places, we shall consider both the cases separately until the acceptable solution is attained.

Make assignment in the cell (3, 2) having the element 1 instead of making zero assignment in the cell (5, 2). After making this assignment we observe that no other assignment can be made in the third row and second column. Consequently make assignment in the cell (5, 4) having element 8, instead of zero assignment. Consequently make assignment in the cell (5, 2).

The new assignment plan is shown in the following table:

		To				
		1	2	3	4	5
	1	∞	12	(0)	⊗	∞
	2	11	∞	⊗	∞	(0)
From	3	⊗	(1)	∞	⊗	1
	4	(0)	∞	⊗	∞	9
	5	∞	⊗	∞	(8)	∞

Thus the resulting feasible solution is

$1 \rightarrow 3 \rightarrow 2 \rightarrow 5 \rightarrow 4 \rightarrow 1$ with cost 9.

Again if we make assignment in the cell (3, 5) having the next best element 1 instead of 0 marked in cell (3, 4), then no feasible solution is obtained in terms of zeros or with cost less than 9.

Hence the optimum route is $1 \rightarrow 3 \rightarrow 2 \rightarrow 5 \rightarrow 4 \rightarrow 1$

The corresponding cost = 4 + 5 + 10 + 20 + 10 = 49.

Example 3(a):

Solve the following L.P.P.

$$\text{Max. } Z = 10x_1 + 3x_2 + 6x_3 + 5x_4$$

$$\text{s.t.} \quad x_1 + 2x_2 + x_4 \le 6$$

$$3x_1 + 2x_3 \le 5$$

$$x_2 + 4x_3 + 5x_4 \le 3$$

and $x_1, x_2, x_3, x_4 \geq 0.$

Compute the limits for a_{11} *and* a_{23} *so that the new solution remains optimal feasible solution.*

Solution:

The given L.P.P. in standard form can be written as follows:

Max. $Z = 10x_1 + 3x_2 + 6x_3 + 5x_4 + 0.x_5 + 0.x_6 + 0.\ x_7$

s.t. $x_1 + 2x_2 + 0.x_3 + x_4 + x_5 = 6$

$3x_1 + 0.x_2 + 2.x_3 + 0.x_4 + x_6 = 5$

$0.x_1 + 1x_2 + 4.x_3 + 5.x_4 + x_7 = 3$

and $x_1, x_2, \ldots, x_7 \geq 0.$

Proceeding as usual the final simplex table is as follows:

		c_j	10	3	6	5	0	0	0
B	c_B	x_B	Y_1	Y_2	Y_3	Y_4	Y_5	Y_6	Y_7
Y_2	3	56/27	0	1	−22/27	0	5/9	−5/27	−1/9
Y_1	10	5/3	1	0	2/3	0	0	1/3	0
Y_4	5	5/27	0	0	26/27	1	−1/9	1/27	2/9
$Z'=c_B\ x_B$=643/27		Δ_j	0	0	−82/27	0	−10/9	−80/27	−7/9

From the above table optimal solution of the given problem is given by

$x_1 = 5/3,\ x_2 = 56/27,\ x_3 = 0,\ x_4 = 5/27$ and $Z = 643/27.$

$\therefore B = (Y_2, Y_1, Y_4) = (y_1, y_2, y_3),\ c_B = (c_{B1}, c_{B2}, c_{B3}) = (3, 10, 5)$

$x_B = (x_{B1}, x_{B2}, x_{B3}) = (56/27, 5/3, 5/27)\ \alpha_5\ \alpha_6\ \alpha_7$

Since the initial (starting) basis matrix was $(Y_5\ Y_6\ Y_7)$:

from above table

$$B^{-1} = (\beta_1, \beta_2, \beta_3) = \begin{bmatrix} 5/9 & 5/27 & 1/9 \\ 0 & 1/3 & 0 \\ -1/9 & 1/27 & 2/9 \end{bmatrix}$$

i.e., $\beta_{11} = 5/9,\ \beta_{21} = 0,\ \beta_{31} = -\ 1/9,$

$\beta_{12} = -\ 5/27,\ \beta_{22} = 1/3,$

$\beta_{32} = 1/27$ and $\beta_{13} = -\ 1/9,$

$\beta_{23} = 0,\ \beta_{33} = 2/9.$

Example 3(b):

Given the matrix of setup costs, show how to sequence the production so as to minimize the setup cost per cycle.

To

From	A_1	A_2	A_3	A_4	A_5
A_1	∞	2	5	7	1
A_2	6	∞	3	8	2
A_3	8	7	∞	4	7
A_4	12	4	6	∞	5
A_5	1	3	2	8	∞

Solution:

Consider the problem as an assignment. Applying the assignment technique, we get the following matrix, showing a solution in terms of marked '( )' zero:

To

From	A_1	A_2	A_3	A_4	A_5
A_1	∞	1	3	6	(0)
A_2	4	∞	(0)	6	⊗
A_3	4	3	∞	(0)	3
A_4	8	(0)	1	∞	1
A_5	(0)	2	⊗	7	∞

The solution to the assignment problem given by above matrix is

$A_1 \to A_2$, $A_5 \to A_1$,

$A_2 \to A_3$, $A_3 \to A_4$,

$A_4 \to A_2$.

This solution indicates to produce the products A_1, then A_5 and then again A_1, without producing the products A_2, A_3 and A_4, which violates the additional restriction of prodding each product once and only once before returning to the first product. So this is not a solution of the travelling salesman problem.

Now we try to find the next best solution which also satisfies the additional restriction. The next minimum element (non-zero) in the matrix is 1. We try to bring 1 in the solution. The cost 1 also occurs at three places.

Start by making unity-assignment in the cell (1, 2) instead of zero assignment in the cell (1, 5). Then no other assignment can be made in the first row and the second column. The best solution of the problem lies in the marked '( )' elements as shown in the table.

		A_1	A_2	A_3	A_4	A_5
	A_1	∞	(1)	3	6	⊗
	A_2	4	∞	(0)	6	⊗
From	A_3	4	3	∞	(0)	3
	A_4	8	⊗	1	∞	(1)
	A_5	(0)	2	⊗	7	∞

Thus, the required solution of the problem is

$A_1 \to A_2 \to A_3 \to A_4 \to A_5 \to A_1$.

For this solution, the cost in the reduced matrix is 2.

On the other hand if we select the element 1 in the cell (4, 3) in the solution, then no feasible solution is available in terms of zero or for which the reduced matrix gives the minimum cost less than 2.

Hence, the most suitable sequence is

$$A_1 \to A_2 \to A_3 \to A_4 \to A_5 \to A_1.$$

The minimum setup cost = 2 + 3 + 4 + 5 + 1 = 15.

Example 4:

Find an optimal solution to the following L.P. problem

$$\text{Max. } Z = 3x_1 + 5x_2$$

$$\text{s.t.} \quad x_1 \le 4$$

$$x_2 \le 6$$

$$3x_1 + 2x_2 \le 18 \text{ and } x_1, x_2 \ge 0,$$

what happens to this optimal solution if the objective is changed to

$$Z^* = 3x_1 + x_2.$$

Solution:

Introducing the slack variables x_3, x_4 and x_5 the given problem in standard form for simplex method is as follows:

$$\text{Max. } Z = 3x_1 + 5x_2 + 0.x_3 + 0.x_4 + 0.x_5$$

$$\text{s.t} \quad 1.x_1 + x_3 = 4$$

$$0.x_1 + 1.x_2 + x_4 = 6$$

$$3x_1 + 2x_2 + x_5 = 18$$

$$\text{and} \quad x_1, x_2, \ldots, x_5 \ge 0.$$

Taking $x_1 = 0 = x_2$,
we get $x_3 = 4$, $x_4 = 6$,
$x_5 = 18$, which is the starting B.F.S.

Proceeding as usual we get the following simplex table.

B	c_B	c_j / x_B	3 Y_1	5 Y_2	0 Y_3	0 Y_4	0 Y_5	Min. ratio x_B/Y_2
Y_3	0	4	1	0	1	0	0	–
Y_4	0	6	0	(1)	0	1	0	6/1 ←
Y_5	0	18	3	2	0	0	1	18/2
Z = 0		Δ_j	3	5 ↑	0	0 ↓	0	x_B/Y_1
Y_3	0	4	1	0	1	0	0	4/1
Y_2	5	6	0	1	0	1	0	–
Y_5	0	6	(3)	0	0	–2	1	6/3 ←
Z = 30		Δ_j	3 ↑	0	0	–5	0 ↓	
Y_3	0	2	0	0	1	2/3	–1/3	
Y_2	5	6	0	1	0	1	0	
Y_1	3	2	1	0	0	–2/3	1/3	
Z = 36		Δ_j	0	0	0	–3	–1	

From the above table the

$\therefore$ Optimal solution of the given L.P.P. is

$$x_1 = 2, x_2 = 6$$

and Max. $Z = 36$.

Now the objective function is changed to $Z^* = 3x_1 + x_2$

i.e., c_2 is changed to 1 keeping c_1 fixed.

To find variation in c_2. Here $c_2 \in c_B$, and from last simplex table

$$c_B = (c_{B1}, c_{B2}, c_{B3}) = (0, 5, 3) = (c_3, c_2, c_1)$$

$\therefore c_2 = c_{Bk} = c_{B2} = 5$.

$\therefore$ Range of Δc_{B2} is given by

$$\underset{y_{2j}>0}{\text{Max}}\left[\frac{c_j - Z_j(=\Delta_j)}{y_{2j}}\right] \le \Delta c_{B2} \le \underset{y_{2j}<0}{\text{Min}}\left[\frac{c_j - Z_j(=\Delta_j)}{y_{2j}}\right]$$

for all j not in the basis.

$$\Rightarrow \text{Max.}\left[\frac{\Delta_4}{y_{24}}\right] \le \Delta c_{B2} < \infty$$

$\Rightarrow -3 \le \Delta c_{B2} < \infty$

$\therefore 5 - 3 \le c_{B2} < 5 + \infty$ or $2 \le c_2 < \infty$

$\therefore$ If c_2 is changed to 1, then the optimal solution given in the above table does not remain optimal.

To find the new optimal solution. When Z is changed to Z*.

If the objective function Z is changed to $Z^* = 3x_1 + x_2$ then c_B in the last simplex becomes (0, 1, 3).

$\therefore \Delta_1 = 0\ \Delta_2 = \Delta_3,$

$\Delta_4 = c_4 - c_B Y_4 = 1,$

$\Delta_5 = c_5 - c_B Y_5 = -1.$

Thus from the last of the above table, we get the new table as follows:

B	c_B	c_j / x_B	3 / Y_1	1 / Y_2	0 / Y_3	0 / Y_4	0 / Y_5	Min. ratio x_B/Y_4
Y_3	0	2	0	0	1	(2/3)	–1/3	2/(2/3) ←
Y_2	1	6	0	1	0	1	0	6/1
Y_1	3	2	1	0	0	–2/3	1/3	
$Z^* = 12$	Δ_j	0	0	0	1 ↓	–1 ↑		
Y_4	0	3	0	0	3/2	1	–1/2	
Y_2	1	3	0	1	–3/2	0	1/2	
Y_1	3	4	1	0	1	0	0	
$Z^* = 15$	Δ_j	0	0	–3/2	0	–1/2		

$\therefore$ The solution of the new L.P.P. with changed objective function is

$x_1 = 4,\ x_2 = 3$

and Max. $Z^* = 15$.

Example 5(a):

Solve the following LP problem

Maximize $Z = -x_1 + 3x_3 - 2x_3$

$$3x_1 - x_2 + 2x_3 \leq 7$$

$$-2x_1 + 4x_4 \leq 12$$

$$-4x_1 + 3x_2 + 8x_3 \leq 10$$

and $x_1, x_2, x_3 \geq 0.$

Discuss the effect of the following changes in the optimal solution.

(a) *Determine the range for discrete changes in the coefficients* a_{13} *and* a_{23} *consistent with the optimal solution of the given LP problem.*

(b) 'x_1'*-column in the problem is changed from* $[3,-2,-4]^T$ *to* $[3, 2, -4]^T$.

(c) 'x_3'*-column in the problem is changed from* $[2, 0, 8]^T$ *to* $[3, 1, 6]^T$.

Solution:

The given LP problem in its standard form can be stated as follows:

Maximize $Z = -x_1 + 3x_2 - 2x_3 + 0.s_1 + 0.s_2 + 0.s_3$

subject to the constraints

$$3x_1 - x_2 + 2x_3 + s_1 = 7$$

$$-2x_1 + 4x_2 \quad + s_2 = 12$$

$$-4x_1 + 3x_2 + 8x_3 + s_3 = 10$$

and $x_1, x_2, x_3, s_1, s_2, s_3 \geq 0$

Applying the simplex (Big-M) method, the optimal solution so obtained is shown in Table 2.2.

Table 2.2 : Optimal Solution

		$c_j \rightarrow$	*–1*	*3*	*–2*	*0*	*0*	*0*
C_B	***Variations in Basis***	***Solution Values***	x_1	x_2	x_3	s_1	s_2	s_3
	B	$b\ (= x_B)$						
–1	x_1	4	1	0	4/5	2/5	1/10	0
3	x_2	5	0	1	2/4	1/5	3/10	0
0	s_3	11	0	0	10	1	–1/2	1
Z = 11		z_j	–1	3	2/5	1/5	4/5	0
		$c_j - z_j$	0	0	–12/5	–1/5	–4/5	0

The optimal basic feasible solution shown in Table 3.2 is:

$x_1 = 4,\ x_2 = 5,\ x_3 = 0$

and Max Z = 11.

In Table 3.2, the inverse of basis matrix, B is

$$B^{-1} = \begin{bmatrix} 2/5 & 1/10 & 0 \\ 1/5 & 3/10 & 0 \\ 1 & -1/2 & 1 \end{bmatrix} = [\beta_1, \beta_2, \beta_3]$$

Thus, we have $c_B\beta_1 = [-1, 3, 0]\ [2/5, 1/5, 1]^T$

$= -1(2/5) + 3(1/5) + 0(1) = 1/5$

$c_B\beta_2 = -1(1/10) + 3(3/10) + 0(-1/2) = 8/10$

$c_B\beta_2 = -1(0) + 3(0) + 0(1) = 0.$

Since variables x_1, x_2 and x_3 are in the basis (see column B of Table 3.1), therefore any discrete change in coefficients, belonging to any of these column vectors may affect both feasibility as well as optimality of the original optimal basic feasible solution, whereas any discrete change in the non-basic variables (i.e. x_5, s_1 and s_2) column vectors may affect only optimality condition.

(i) Ranges for discrete change in coefficients a_{13} and a_{23} in the x_3-column vector of Table 3.2 are computed as:

$$\text{Max}\left\{\frac{c_3 - z_3}{c_B\,\beta_1}\right\} = \text{Max}\left\{\frac{-12/5}{1/5}\right\} \le \Delta a_{13} \text{ or } \Delta a_{13} \ge -12$$

and

$$\text{Max}\left\{\frac{c_3 - z_3}{c_B\,\beta_2}\right\} = \text{Max}\left\{\frac{-12/5}{8/10}\right\} \le \Delta a_{23} \text{ or } \Delta a_{23} \ge -3$$

(ii) To measure the change Δa_{21} in the coefficient a_{21} (= 2) in the first column of variable x_1 in the second constraint of the original set of constraints, we need to check both feasibility as well as optimality conditions because variable x_1 is the basic variable as shown in Table 3.2.

(a) *Feasibility condition.* For i = 2 (constraint), p = 1 (column), and k = 2, 3 (columns of B^{-1}), we have

For k = 2 $\quad x_{Bk}\,\beta_{pi} - x_{Bp}\,\beta_{12} = x_{B2}\,\beta_{12} - x_{B1}\,\beta_{22}$

$= 5(1/10) - 4(3/10) = -7/10$

For k = 3 $\quad x_{B3}\,\beta_{12} - x_{B1}\,\beta_{32} = 11(1/10) - 4\ (-12) = 32/10$

Hence, the range to maintain feasibility of the existing optimal solution is

$$\frac{-5}{\frac{31}{10}} \leq \Delta a_{12} \leq \frac{-5}{\frac{-7}{10}}$$

$$2-(50/31) \leq a_{21} \leq 2+(50/7) \text{ or } -12/31 \leq a_{21} \leq 64/7$$

(b) Optimality condition

$$(c_3-z_3)\beta_{12}-y_{13}c_B\beta_2=-\frac{12}{5}\left(\frac{1}{10}\right)-\frac{4}{5}\left(\frac{8}{10}\right)=-\frac{44}{50}$$

$$(c_4-z_4)\beta_{12}-y_{14}c_B\beta_2=-\frac{1}{5}\left(\frac{1}{10}\right)-\frac{2}{5}\left(\frac{8}{10}\right)=-\frac{17}{50}$$

$$(c_5-z_5)\beta_{12}-y_{14}c_B\beta_2=-\frac{4}{5}\left(\frac{1}{10}\right)-\frac{1}{10}\left(\frac{8}{10}\right)=-\frac{16}{100}$$

Hence, the range to maintain optimality of the existing optimal solution is

$$-\infty \leq \Delta a_{12} \leq \text{Min}\left\{\frac{-12/5}{-44/50};\frac{-1/5}{-17/50};\frac{-4/5}{-16/100}\right\}$$

$$-\infty \leq \Delta a_{21} \leq 10/17 \text{ or } -\infty \leq a_{21} \leq 44/17$$

(c) Suppose column vector 33 (x_3-column in Table 3.2) of original LP model is changed from $[2, 0, 8]^T$ to $[3, 1, 6]^T$. Then new value of $c_3 - z^*_3$ for this column is:

$$B^{-1}a_3^* \begin{bmatrix} 2/5 & 1/10 & 0 \\ 1/5 & 3/10 & 0 \\ 1 & -1/2 & 1 \end{bmatrix}\begin{bmatrix} 3 \\ 1 \\ 0 \end{bmatrix}=\begin{bmatrix} 13/10 \\ 9/10 \\ 17/2 \end{bmatrix}$$

$$c_3-z^*_3=c_3\,B^{-1}\,a^*_3=-2-[-1,3,0]\begin{bmatrix} 13/10 \\ 9/10 \\ 17/2 \end{bmatrix}=-\frac{34}{10}$$

Since all entries in the x_3-column are non-zero and so we replace column Xg with new entries in Table 6.10 and proceed to get new optimal solution.

Addition of a New Variable (Column)

Let an extra variable x_{n-1} with coefficient c_{n+1} be added in the system of original constraint. Ax = B, x ≥ 0. Thus, it creates an extra column a_{n+1} the matrix A ofcoeffcients. To see the impact of this addition on the current optimal solution, we compute

$$V_{n+1} = B^{-1}\,a_{n+1}$$

and $\quad c_{n+1} - z_{n+1} = c_{n+1} - c_B y_{n+1}$

Two cases of the maximization LP model may arise:

1. If $c_{n+1} - z_{n+1} \leq 0$, then $x_B = 0$, and hence current solution remains optimal.
2. If $c_{n+1} - z_{n+1} > 0$, then the current optimal solution can be improved by introducing a new column a_{n+1} into the basis to find the new optimal solution.

Addition of a New Constraint (Row)

After solving an LP model, the decision-maker may recall that a particular resource constraint was overlooked in the model formulation or perhaps he may desire to know the effect of adding a new resource to enhance the objective function value. Addition of a constraint in the existing constraints will cause a simultaneous change in the objective function coefficients (c_j), as well as coefficients a_{ij} of a corresponding non-basic variable. Thus, it will affect only the optimality of the problem. This means that the new variable should enter into the basis only if it improves the value of the objective function.

Suppose that a new constraint

$$a_{m+1,1}, x_1 + a_{m+1,2} x_2 = ...; + a_{m+1,n} x_n \leq b_{m+1}$$

is added to the system of original constraints $Ax = b$, $x \geq 0$; where b_{m+1}, is positive, zero or negative. Then, two cases may arise.

1. The optimal solution (x_B) of the original problem satisfies the new constraint. If it is so, the solution remains feasible as well as optimal, because the new constraint either reduces or leaves the feasible region of the given LP problem unchanged.
2. The optimal solution (x_B) of the original problem does not satisfy the new constraint. Then the optimal solution to the modified LP problem is re-obtained. Let B be the basis matrix for the original problem and B_1 be the basis matrix for the new problem with $m + 1$ constraints. That is, matrix B_1 of order $(m + 1)$ is given by

$$B_1 = \begin{bmatrix} B & 0 \\ \alpha & 1 \end{bmatrix}$$

where the second column of B_1 corresponds to slack, surplus or artificial variables added to the new constraint and $\alpha = (a_{m+1,1}, a_{m+1,2}, \ldots a_{m+1,n})$ a is a row vector containing the coefficients in the new constraint and corresponds to variables in the optimal basis. In order to prove that the new solution

$$x_B^* = \begin{bmatrix} X_B \\ s \end{bmatrix} \quad s = \text{slack variable}$$

is a basic feasible solution to the new LP problem we shall compute the inverse of B_1 using partitioned methods as given below:

$$x_1^{-1} = \begin{bmatrix} B^{-1} & 0 \\ -\alpha B^{-1} & 1 \end{bmatrix}$$

Since each column vector in the new LP problem is given by $a_j^* = (a_j, a_{m+1,j})$, the new columns y_j^*. are given by

$$y_j^* B^{-1} a_j^* = \begin{bmatrix} B^{-1} & 0 \\ -\alpha B^{-1} & 1 \end{bmatrix} \begin{bmatrix} a_j \\ a_{m+1,j} \end{bmatrix}$$

$$= \begin{bmatrix} B^{-1}a_j \\ -\alpha B^{-1}a_j + a_{m+1,j} \end{bmatrix} = \begin{bmatrix} y_j \\ a_{m+1,j} - \alpha_j \end{bmatrix}$$

The entries in the $c_j - z^*_j$ row of the simplex table for any non-basic variable x. in the new problem are computed as follows:

$$c_j - z_j^* = c_j - z_j^* y_j^* = c_j - [c_B, c_{Bm+1}] \begin{bmatrix} y_j \\ a_{m+1,j} - \alpha y_j \end{bmatrix}$$

$$= c_j - (c_B y_j + c_{Bm+1}\, a_{m+1} - c_{Bm+1}\, \alpha_{yj})$$

where cB_{m+1}, is the coefficient associated with the new variable introduced in the basis of the new LP problem.

If the new variable introduced in the basis of the new LP problem is slack or surplus variable, then $c_{Bm+1} = 0$. Hence, we have

$$c_j - z^*_j = c_j - c_B y_j = c_j - z_j$$

That is, the entries in the $c_j - z_j$ row are the same for the initial and new LP problem. The value of objective function is given by

$$Z^* = [c_B, 0] \begin{bmatrix} x_B \\ s \end{bmatrix} = c_B x_B = Z$$

This shows that the optimal simplex table of the original problem remains unchanged even after adding the new constraint. However, the slack or surplus variable appears with negative value because the optimal solution of the original problem does not satisfy the new constraints. Thus, the dual simplex method may be used to get an optimal solution.

Remarks : If the new constraint added is an equation and an artificial variable appears in the basis of the new problem, then two cases may arise:

1. If an artificial variable appears in the basis at negative value, a zero cost may be assigned to it. Apply the dual simplex method to obtain an optimal solution.
2. If the artificial variable appears in the basis at a positive value, then – M cost may be assigned to it. Apply the usual simplex method to obtain an optimal solution.

Example 5(b):

Solve the following L.P.P.

$$Max.\ Z = 3x_1 + 5x_2$$

$$s.t.\quad x_1 + x_3 = 4$$

$$3x_1 + 2x_2 + x_4 = 18$$

$$and\quad x_1, x_2, x_3, x_4 \geq 0.$$

If a new variable x_5 is introduced in the above L.P.P. with price 7, then we have the following problem:

$$Max.\ Z' = 3x_1 + 5x_2 + 7x_5$$

$$s.t.\quad x_1 + x_3 + x_5 = 4$$

$$3x_1 + 2x_2 + x_4 + 2x_5 = 18$$

$$and\quad x_1, x_2, x_3, x_4, x_5 \geq 0.$$

Find the solution of the new L.P.P.

Solution:

Proceeding as usual the successive simplex tables for the original L.P.P. are as follows:

Simplex Table 1

		c_j	3	5	0	0	Mini Ratio
B	c_B	x_B	Y_1	Y_2	Y_3	Y_4	x_B/Y_2
Y_3	0	4	1	0	1	0	–
Y_4	0	18	3	(2)	0	1	18/2 ←
$Z=c_B.x_B=0$		Δ_j	3	5 ↑	0	0 ↓	
Y_3	0	4	1	0	1	0	
Y_2	5	9	3/2	1	0	1/2	
$Z=c_B.x_B=45$		Δ_j	–9/2	0	0	–5/2	

∴ Optimal solution of the given L.P.P. is

$x_1 = 0,\ x_2 = 9,\ x_3 = 4,$

$x_4 = 0,$ Max. $Z = 45.$

From the final table the solution of the dual of the given L.P.P. is

$w_1 = 0,\ w_2 = 5/2,$

Min. $Z_D = 45.$

Revised L.P.P. The dual of the new L.P.P. is same as that of the original L.P.P. with one more constraint $w_1 + 2w_2 \geq 7$ which corresponds to the new variable x_5.

The optimal solution of the dual of the original L.P.P. is $w_1 = 0$, $w_2 = 5/2$ which does not satisfy this constraint. Thus we see that the optimal solution of the dual of the given L.P.P. is not optimal for the revised problem and can be improved by introducing $\alpha_5 = [1, 2]$, (column of A corresponding to the new variable introduced) in the basis.

Thus, consider one more column Y_5 in the above table.

Since $B^{-1} = \begin{bmatrix} 1 & 0 \\ 0 & 1/2 \end{bmatrix}$

$$\therefore \quad Y_5 = B^{-1} \alpha_5 = \begin{bmatrix} 1 & 0 \\ 0 & 1/2 \end{bmatrix} \cdot \begin{bmatrix} 1 \\ 2 \end{bmatrix} = \begin{bmatrix} 1 \\ 1 \end{bmatrix}$$

$$\Delta_5 = c_5 - c_B Y_5 = 7 - (0, 5), (1, 1) = 2.$$

We get the following simplex table for revised L.P.P. as

Simplex Table 2

B	c_B	c_j / x_B	3 / Y_1	5 / Y_2	0 / Y_3	0 / Y_4	7 / Y_5	Mini Ratio x_B/Y_5
Y_3	0	4	1	0	1	0	(1)	4/1 ←
Y_2	5	9	3/2	1	0	1/2	1	9/1
$Z'=c_B\ x_B=45$		Δ_j	–9/2	0	0 ↓	–5/2 ↑	2	
Y_5	7	4	1	0	1	0	1	
Y_2	5	5	1/2	1	–1	1/2	0	
$Z'=c_B\ x_B=53$		Δ_j	–13/2	0	–2	–5/2	0	

∴ Optimal solution of revised L.P.P. is

$x_1 = 0, x_2 = 5, x_3 = 0,$

$x_4 = 0, x_5 = 4.$

and Max. Z' = 53.

Example 5(c):

Find an optimal solution to the following L.P.P.

Max. $Z = 15x_1 + 45x_2$

s.t. $x_2 \leq 50$

$x_1 + 1.6x_2 \leq 240$

$5x_1 + 2.9x_2 \leq 162$

and $x_1, x_2 \ ^3\ 0.$

If Max. $Z = \sum c_i x_i$ $(i = 1, 2)$ *and* c_2 *is kept fixed at 45, find how much can* c_1 *be changed without affecting the above optimal solution.*

Solution:

Introducing the slack variable x_3, x_4 and x_5, the given problem in standard form for simplex method is as follows:

Max. $Z = 15x_1 + 45x_2 + 0.x_3 + 0.x_4 + 0.x_5$

s.t. $0.x_1 + x_2 + x_3 = 50$

$x_1 + 1.6x_2 + x_4 = 240$

$5x_1 + 2.0x_2 + x_5 = 162$

and $x_1, x_2, x_3, x_4, x_5 \geq 0.$

Taking $x_1 = 0 = x_2$, $x_3 = 50$, $x_4 = 240$, $x_5 = 162$, which is the starting B.F.S.

Solving the given L.P.P. by simple method, we get the following simplex table.

		c_j	15	45	0	0	0	Min. ratio x_B/Y_2
B	c_B	x_B	Y_1	Y_2	Y_3	Y_4	Y_5	
Y_3	0	50	0	(1)	1	0	0	50/1 ←
Y_4	0	240	1	1.6	0	1	0	240/1.6
Y_5	0	162	.5	2	0	0	1	162/2
$Z = c_B x_B = 0$		Δ_j	15	45 ↑	0 ↓	0	0	x_B/Y_1

Y_2	45	50	0	1	1	0	0	–
Y_4	0	160	1	0	–1.6	1	0	160/1
Y_5	0	62	(.5)	0	–2	0	1	26/.5
								←
Z = 2250	Δ_j	15	0	–45	0	1	x_B/Y_3	
		↑				↓		
Y_2	45	50	0	1	1	0	0	50/1
Y_4	0	36	0	0	(2.4)	1	–2	36/2.4
								←
Y_1	15	124	1	0	–4	0	2	–
Z = 4110	Δ_j	0	0	15	0	–30		
				↑	↓			
Y_2	45	35	0	1	0	–5/12	5/6	
Y_3	0	15	0	0	1	5/12	–5/6	
Y_1	15	184	1	0	0	5/3	–4/3	
Z = 4335	Δ_j	0	0	0	–25/4	–35/2		

From the table we set the optimal solution of the given L.P.P is

$x_1 = 184,\ x_2 = 35,$ Max. $Z = 4335.$

To find variation in c_1. Here $c_1 \in c_B$.

From the above simplex table, we have

$c_B = (c_{B1}, c_{B2}, c_{B3}) = (45, 0, 15) = (c_2, c_3, c_1)$

$\therefore c_1 = c_{Bk} = c_{B3} = 15.$

$\therefore$ We have to find the range of variation in $c_1 \in c_B$.

Here we want to find the range $\Delta\, c_{B3}$ (change in c_{B3}), which is given by (3), 2.

$$\underset{y_{kj}>0}{\text{Max}}\left[\frac{c_j - Z_j(=\Delta_j)}{y_{kj}}\right] \le \Delta c_{Bk} \le \underset{y_{kj}<0}{\text{Min}}\left[\frac{c_j - Z_j(=\Delta_j)}{y_{kj}}\right]$$

Here k = 3.

Here $y_{kj} = y_{3j} = (1, 0, 0, -4/3)$, out of which y_{31}, y_{32}, y_{33} are not to be taken j = 1, 2, 3 correspond to $\alpha_1, \alpha_2, \alpha_3$ which are in the basis. Thus here we shall consider y_{34} and y_{35} only.

Now $y_{34} = 5/3 > 0$

and $y_{35} = -4/3 < 0$.

$\therefore \Delta c_{B3}$ is given by

$$\text{Max.}\left[\frac{\Delta_4}{y_{34}}\right] \leq \Delta c_{B3} \leq \text{Min.}\left[\frac{\Delta_5}{y_{35}}\right]$$

or $$\text{Max.}\left[\frac{-25/4}{5/3}\right] \leq \Delta c_{B3} \leq \text{Min.}\left[\frac{-35/2}{-4/3}\right]$$

or $-15/4 \leq \Delta c_{B3} \leq 105/8$.

$\therefore$ Range of variation of $c_{B3} = c_1$ is given by

$c_{B3} + (-15/4) \leq c_1 \leq c_{B3} + 105/8$

or $15 + (-15/4) \leq c_1 \leq 15 + 105/8$

or $45/4 \leq c_1 \leq 225/8$.

If we take $c_1 = 45/4$, then proceeding as in above table, we do not get the same optimal solution.

$\therefore$ The variation in c_1 without affecting the above optimal solution is given by $45/4 < c_1 \leq 225/8$.

EXERCISES

1. (a) Determine optimal solution to the problem

   (b) Determine the effect of discrete changes in those components of the cost vector which corresponds to the basic variable.

2. A firm uses three machines in the manufacture of three products. Each unit of product I requires 3 hours on machine 1, 2 hours on .machine 2 and 1 hour on machine 3. Each unit of product II requires 4 hours on machine 1, 1 hour on machine 2 and 3 hours on machine 3. Each unit of product III requires 2 hours on machine 1, 2 hours on machine 2, and 2 hours on machine 3. The contribution margin of the three products is Rs. 30, Rs. 40 and Rs. 35 per unit, respectively. Available for scheduling are 90 hours of machine 1 time, 54 hours of machine 2 time, and 93 hours of machine 3 time.

   (a) What is the optimal production schedule for the firm?

   (b) What is the marginal value of an additional hour of time on machine?

   (c) What is the opportunity cost associated with product I? What interpretation should be given to this opportunity cost?

(d) Suppose that the contribution margin for product I is increased from Rs. 30 to Rs. 43, would this change the optimal production plan? Give reasons.

3. Given the LP problem

$$\text{Max } Z = 3x_1 + 5x_2$$

$$\text{subject to} \quad 3x_1 + 2x_3 \leq 18$$

$$x_1 \leq 4$$

$$x_2 \leq 6$$

$$\text{and} \quad x_1, x_2 \geq 0$$

Discuss the effect on the optimality of the solution when the objective function is changed to $3x_1 + x_2$.

4. A stainless steel utensil manufacturer makes three types of items. The restrictions, profits and requirements are tabulated below:

Utensil Type	*I*	*II*	*III*
Raw material requirement (kg per unit)	6	3	5
Welding and finishing time (hours per unit)	3	4	5
Profit per unit (Rs.)	3	1	4

If stainless steel (raw material) availability is 25 kg and welding and finishing time available is 20 hours per day, then the optimum product mix problem is expressed as:

$$\text{Max } Z = 3x_1 + x_2 + 4x_3$$

$$\text{subject to} \quad 6x_1 + 3x_2 + 5x_3 \leq 25 \quad \text{(raw material restriction)}$$

$$3x_1 + 4x_2 + 5x_3 \leq 20 \quad \text{(time restriction)}$$

$$\text{and} \quad x_1, x_2, x_3 \geq 0$$

where x_j ($j = 1, 2, 3$) is the number of units of the jth type of the item to be produced. Get the optimal simplex table and encircle the appropriate answer to the following questions.

(a) The second type of utensil would change the current optimal basis if its profit per unit is: ≥ 1; ≥ 1.02; ≥ 1.06; ≥ 2; ≥ 3.

(b) The simplex multiplier associated with the machine time restriction of 20 hours is (–3/5). Thus, the multiplier remains unchanged for the upper limit on the machine time availability of 25; 27.5; 35; 42.5; 32.5.

5. Given the LP problem

$$\text{Max } Z = x_1 + 2x_2 + x_3$$

subject to $3x_1 + x_2 - x_3 \leq 10$

$-x_1 + x_2 + x_3 \leq 6$

$x_2 + x_3 \leq 4$

and $x_1, x_2, x_3 \geq 10.$

(a) Determine the sensitivity limits for unit profits within which the current optimal solution will remain unchanged.

(b) What is the total profit when each table yields a profit of Rs. 180 and each chair yields a profit of Rs. 100.

6. Given, that the problem: Max Z = c x such that A x = b, x ≥ 0 has an optimal solution, can one obtain a linear programming problem which has an unbounded solution changing b alone.

7. Find the limits of variation of element a_{ik}, so that optimal feasible solution of Ax = b, x ≥ 0, Max Z = ex remains optimal feasible solution when (i) $a_k \in B$, (ii) $a_k \notin B$.

8. Consider the following LP problem

   Max $Z = 4x_1 + 6x_2$

   subject to $x_1 + 2x_2 \leq 8$

   $6x_1 + 4x_2 + \leq 24$

   and $x_1, x_2 \geq 0$

   (a) What is the optimal solution?

   (b) If the first constraint is altered as: $x_1 + 3x_2 \leq 8$, does the optimal solution change?

9. Discuss the role of sensitivity analysis in linear programming. Under what circumstances is it needed, and under what conditions do you think it is not necessary?

10. A company makes two products: A and B. The production of both products requires processing time' in two departments I and II. The hourly capacity of I and II, unit profits for products: A and B and the processing time requirements in I and II are given in the following table:

Department	***Product***		***Capacity (Hours)***
	A	***B***	
I	1	2	32
II	0	1	8
Unit profit (Rs.)	200	300	

Now the company is considering the addition of a new product C to its line. Product C requires one hour each of department I and II. What must be product C's unit profit in order to profitably add it to the firm's product line?

11. Find the optimal solution to the LP problem

Max $Z = 15x_1 + 45x_2$

subject to
$$x_1 + 16x_2 \leq 250$$
$$5x_1 + 2x_2 \leq 162$$
$$x_2 \leq 50$$
and
$$x_1, x_2 \geq 0$$

If Max $Z = \Sigma c_j x_j, j = 1, 2$ and c_2 is kept fixed at 45, determine how much can c_1 be changed without affecting the optimal solution of the problem.

12. The increase in the objective function for each unit availability of machine time higher than the upper limit indicated in part (b) is ... (show calculations).

13. The profit of the third type of utensil is Rs. 4 per unit. The lower unit on its profitability such that the current basis is still optimal is 4; 3; 2.5; 2; < 2.

14. Discuss the effect of discrete changes in the parameter b_1 (i = 1, 2, 3) for the LP problem.

Max $Z = 3x_1 + 4x_2 + x_3 + x_4$

subject to
$$8x_1 + 3x_2 + 4x_3 + x_4 \leq 7$$
$$2x_1 + 6x_2 + x_3 + 5x_4 \leq 3$$
$$x_1 + 4x_2 + 5x_3 + 2x_4 \leq 8$$
and
$$x_1, x_2, x_3, x_4 \geq 0$$

15. Given the LP problem

subject to
$$3x_1 + x_2 - 4x_3 \leq 10$$
$$-x_1 + 4x_2 + x_3 \geq 6$$
$$x_2 + x_3 \leq 4$$
and
$$x_1, x_2, x_3 \geq 0.$$

Determine the range for discrete changes in the resource values $b_1 = 10$ and $b_2 = 6$ of the LP. model so as to maintain optimality of the current solution.

16. A pig farmer is attempting to analyse his feeding operation. The minimum daily requirement of the three nutritional elements for the pigs and the number of units of each of these nutritional elements in two feeds is given in the following table:

Required Nutritional Element	*Units of Nutritional Elements (in kg)*		*Minimum Requirement*
	Food 1	*Food 2*	
A	20	30	200
B	40	25	350
C	30	45	430
Cost per kg	5	3	

   (a) Formulate and solve this problem as an LP model.

   (b) Assume that the farmer can purchase a third feed at a cost of Rs. 2 per kg, which will provide 35 units units of nutrient A, 30 units of nutrient B and 50 units of nutrient C. Would this change the optimal mix of feeds? If yes, how?

17. (a) Explain how a change in a input-output coefficient can affect a problem's optimal solution?

   (b) How can a change in resource availability affect a solution?

18. Consider the LP problem; Max Z = ex; subject to Ax = b and $x \geq 0$, where c, $x^T \in E^n$, $b^T \in E^m$ and A is $m \times n$ coefficients matrix. Determine how much the components of the cost vector c can be changed without affecting the optimal solution of the LP problem.

19. The following table gives the optimal solution to a L.P.P.

   Max. $Z = 3x_1 + 5x_2 + 4x_3$,

   s.t. $2x_1 + 3x_2 \leq 8$, $2x_2 + 5x_3 \leq 10$,

   $3x_1 + 2x_2 + 4x_3 \leq 15$. $x_1, x_2, x_3 \geq 0$.

   For the above L.P.P. calculate the following:

   (i) How much c_3 and c_4 can be increased before the present basic solution will no longer be optimal? Also, find the change in the value of the objective function if possible.

   (ii) How much b_2 can be changed maintaining the feasibility of the solution?

20. For the L.P.P. Max. $Z = 5x_1 + 3x_2$

   Subject to, $3x_1 + 5x_2 \leq 15$

   $5x_1 + 2x_2 \leq 10$

   and $x_1, x_2 \geq 0$.

Find an optimal solution. Hence, find how far the component c_1 of the vector c of the function $Z = cx$ can be increased without destroying the optimality of the solution.

21. Solve the following L.P.P.

Max. $Z = 3x_1 + 5x_2$

s.t. $x_1 + x_3 = 4$

$3x_1 + 2x_2 + x_4 = 18$

and $x_1, x_2, x_3, x_4 \,{}^3\, 0.$

Would there be any change in the optimal solution if the additional constraint (1) x_2 £ 10 or (ii) x_2 £ 6 is added to the above L.P.P. In case the optimal solution change find the new optimal feasible solution.

22. (i) Discuss the effect of discrete changes in the requirements (on the right side of the inequalities) for the following L.P.P.

Max. $Z = 3x_1 + 4x_2 + x_3 + 7x_4$

subject to

$8x_1 + 3x_2 + 4x_3 + x_4 \leq 7$

$2x_1 + 6x_2 + x_3 + 5x_4 \leq 3$

$x_1 + 4x_2 + 5x_3 + 2x_4 \leq 8,$

$x_1\ x_2, x_3, x_4 \geq 0.$

(ii) Discuss the effect of discrete changes in c on the optimality of an optimum basis feasible solution to the above L.P.P.

23. Solve the following L.P.P. and find how much can c_1 be changed without affecting the optimal conditions.

Max. $Z = 15x_1 + 45x_2 = c_1\, x_1 + c_2\, x_2$

s.t. $x_1 + 1.6x_2 \leq 240$

$.5x_1 + 2x_2 + x_3 = 162,$

$x_2 + x_4 = 50,$

$x_1, x_2, x_3, x_4 \geq 0.$

24. A company sells two different products: A and B. The selling price and incremental cost information is as follows:

	Product A	*Product B*
Selling price (Rs.)	60	40
Incremental cost (Rs.).	30	10
Incremental profit (Rs.)	30	30

The two products are produced in a common production process and are sold in two different markets. The production process has a capacity of 30,000 labour hours. It takes three hours to produce a unit of A and one hour to produce a unit ofB. The market has been surveyed and company officials feel that the maximum number of units of A that can be sold is 8,000; and that of B is 12,000 units.

(a) Find the optimal product mix.

(b) Suppose maximum number of units of A and B that can be sold is actually 9,000 units and 13,000 units, respectively instead of as given in the problem, what effect does this have on the solution? What is the effect on the profit? What is the shadow price for the constraint on the sales limit for both these products.

(c) Suppose there are 31,000 labour hours available instead of 30,000 as in the base case, what effect does this have on the solution? What is the effect on the profit? What is the shadow price for the constraint on the number of labour hours.

25. Write a short note on sensitivity analysis.

26. Given that the problem: Max. Z = cx such that Ax = b, x ≥ 0, has an optimal solution; can one obtain a linear programming problem which has an unbounded solution changing b along?

27. An organization can produce a particular component for passenger cars, jeeps and trucks. The production of the component requires utilization of sheet metal working and painting facilities, the details of which are given below.

Resource	*Consumption to Produce a Unit for*			*Availability*
	Passenger car	*Jeep*	*Truck*	
Sheet metal working	0.25 hr	1 hr	0.5 hr	12 hrs
Painting	0.5 hr	1 hr	2 hr	30 hrs

The profits that can be earned by the three categories of components, i.e. for passenger cars, jeeps and trucks are Rs. 600, Rs. 1400, and Rs. 1300, respectively.

(a) Find the optimal product mix for that organization.

(b) What additional profit would be earned by increasing the availability of:

(i) Sheet metal working shop by an hour only.

(ii) Painting shop by an hour only.

(c) What would be the effect on the profit earned if at least one component for jeep had to be produced?

28. Find an optimal solution to the following L.P.P.

    Max. $Z = 15x_1 + 45x_2$

    s.t. $x_1 + 16x_2 \leq 240$

    $5x_1 + 2x_2 \leq 162$

    $x_2 \leq 50,\ x_1,\ x_2 \geq 0.$

    If max. $Z = \sum c_j x_j$, $j = 1, 2$ and c_2 is fixed at 45, determine how much c_1 can be changed without affecting the above solution.

29. Discuss the changes in the coefficients a_{ij} for the given LP problem: Max $Z = ex$, subject to $Ax = b$, $x \geq 0$.

30. Write a short note on travelling salesman problem.

31. State the travelling salesman problem and formulate it as an assignment problem.

32. How can the travelling select man problem be solved using assignment algorithm.

33. Solve the 'travelling salesman problem' given by the following data:

    $c_{12} = 4$, $c_{13} = 7$,

    $c_{14} = 3$, $c_{23} = 6$,

    $c_{24} = 3$ and $c_{34} = 7$,

    where $c_{ij} = c_{ji}$.

34. Write a short note on problem analysis.

35. What do you understand by the term 'sensitivity analysis'? Discuss the effect of

    (i) variation of c_j,

    (ii) variation of b_i, and

    (iii) addition of a new constraint.

5

Decision Theory, Trees, Games and Investment Analysis

INTRODUCTION

The success or failure that an individual or organization experiences, depends to a large extent on the ability of making appropriate decisions. Making of a decision requires an enumeration of feasible and viable alternatives (courses of action or strategies), the projection of consequences associated with different alternatives, and a measure of effectiveness (or an objective) by which the most preferred alternative is identified. *Decision theory* provides an analytical and systematic approach to the study of decision-making. In other words, decision theory provide a method of natural decision-making wherein data concerning the occurrence of different outcomes (consequences) may be evaluated to enable the decision-maker to identify suitable alternative (or course of action).

Decision models useful in helping decision-makers make the best possible decisions are classified according to the *degree of* certainty. The scale of certainty can range from complete certainty to complete uncertainty. The region which falls between these two extreme points corresponds to the decision making under risk (probabilistic problems).

Irrespective of the type of decision model, there are certain essential characteristics which are common to all as listed below.

Decision Alternatives : There is a finite number of decision alternatives available with the decision-maker at each point in time when a decision is made. The number and type of such alternatives may depend on the previous decisions made and on what has happened subsequent to those decisions. These alternatives are also called courses of action (actions, acts or strategies) and are under control and known to the decision-maker. These may be described numerically such as, stocking 100 units of a particular item, or non-numerically such as, conducting a market survey to know the likely demand of an item.

State of Nature : A possible future condition (consequence or event) resulting from the choice of a decision alternative depends upon certain factors beyond the control of the decision-maker. These factors are called states of nature (future). For example, if the decision is to carry an umbrella or not, the consequence (get wet or do not) depends on what action nature takes.

The states of nature are mutually exclusive and collectively exhaustive with respect to any decision problem. The states of nature may be described numerically such as, demand of 100 units of an item or non-numerically such as, employees strike, etc.

Payoff : A numerical value resulting from each possible combination of alternatives and states of nature is called payoff. The payoff values are always conditional values because of unknown states of nature.

A tabular arrangement of these conditional outcome (payoff) values is known as payoff matrix as shown in Table 5.1.

Table 5.1 : General Form of Payoff Matrix

	Courses of Action (Alternatives)			
States of Nature	S_1	S_2		S_3
N_1	P_{11}	P_{12}	...	P_{1n}
N_2	P_{12}	P_{22}	...	P_{2n}
N_3	P_{m1}	P_{m2}	...	P_{mn}

STEPS IN DECISION THEORY APPROACH

The decision theory approach generally involved four steps. We shall introduce these by considering a manufacturing company that is thinking of several alternative methods to increase its production to meet the increasing market demand.

Steps 1 : List all Viable Alternatives

The first action a decision maker must take is to list all the viable alternatives that can be considered in the decision. For instance, for the company considered above, there may be only following three options:

1. Expand the present plant
2. Construct a new plant
3. Subcontract production for extra demand.

Steps 2 : Identify the Expanded Future Events

The second step of the decision maker is to list all the future events that may occur. Often it is possible to identify most to the events that can occur, the difficulty is to identify which particular event will occur. These future events (not under the control of decision maker) are termed as states of nature in decision theory. Of coarse, out of this list, only one of the events will occur. For the manufacturing company considered above, the greatest uncertainty will be about product demand. The future events related to the demand may be:

1. High demand
2. Moderate demand
3. Low demand
4. No demand.

Steps 3 : Construct a Payoff Table

The decision maker now constructs a payoff table (table representing profit, benefit etc.) for each possible combination of alternative course of action and state of nature.

Steps 4 : Select Optimum Decision Criterion

Finally, the decision maker will choose criterion which results in largest payoff. The criterion may be economic, quantitative or qualitative (*e.g.*, market share, profit, fragrance of a perfume, etc.).

The decision-making process involves the following steps:

(i) Identify and define the problem

(ii) Listing of all possible future events, called states of nature, which can occur in the context of the decision problem. Such events are not under the control of decision-maker because these are erratic in nature.

(iii) Identification of all the courses of action (alternatives or decision choices) which are available to the decision-maker. The decision-maker has control over these courses of action.

(iv) Expressing the payoffs (p_{ij}) resulting from each pair of course of action and state of nature. These payoffs are normally expressed in a monetary value.

(v) Apply an appropriate mathematical decision theory model to select best course of action from the given list on the basis of some criterion (measure of effectiveness) that results in the optimal (desired) payoff.

TYPES OF DECISION MAKING ENVIRONMENTS

Decisions are made based upon the data available about the occurrence of events as well as the decision situation (or environment). There are four types of decision-making environment: *Certainty, uncertainty, risk and conflict.*

Type 1: Decision Making Under Certainty

In this case the decision-maker has the complete knowledge (perfect information) of consequence of every decision choice (course of action or alternative) with certainty. Obviously, he will select an alternative that yields the largest return (payoff) for the known future (state of nature). For example, the decision to purchase either N.S.C. (National Saving Certificate); Indira Vikas Patra, or deposit in N.S.S. (National Saving Scheme) is one in which it is reasonable to assume complete information about the future because there is no doubt that the Indian government will pay the interest when it is due and the principal at maturity. In this decision-model, only one possible state of nature (future) exists.

Type 2: Decision Making Under Risk

In this case the decision-maker has less than complete knowledge with certainty of the consequence of every decision choice (course of action). This means there is more than one state of nature (future) and for which he makes an assumption of the probability with which each state of nature will occur. For example, probability of getting head in the toss of a coin is 0.5.

Type 3: Decision Making Under Uncertainty

In this case the decision-maker is unable to specify the probabilities with which the various states of nature (futures) will occur. Thus, decisions under uncertainty are taken with even less information than decisions under risk. For example, the probability that Mr. X will be the prime minister of the country 15 years from now is not join own.

DECISION MAKING UNDER UNCERTAINTY

In the absence of knowledge about the probability of any state of nature (future) occurring, the decision-maker must arrive at a decision only on the actual conditional payoff values, together with a policy (attitude). There are several different criteria of decision-making in this situation. The criteria that we will discuss in this section include

(i) Maximax or Minimin
(ii) Maximin or Minimax
(iii) Equally likely
(iv) Criterion of realism
(v) Criterion of regret.

Criterion of Optimism (Maximax or Minimin)

In this criterion the decision-maker ensures that he should not miss the opportunity to achieve the largest possible profit (maximax) or lowest possible cost (minimin). Thus, he selects the alternative (decision choice or course of action) that represents the maximum of the maxima (or minimum of the minima) payoffs (consequences or outcomes). The working method is summarized as follows:

(a) Locate the maximum (or minimum) payoff values corresponding to each alternative (or course of action), then

(b) Select an alternative with best anticipated payoff value (maximum for profit and minimum for cost).

Since in this criterion the decision-maker selects an alternative with largest (or lowest) possible payoff value, it is also called an *optimistic decision criterion.*

Criterion of Pessimism (Maximin or Minimax)

In this criterion the decision-maker ensures that he would earn no less (or pay no more) than some specified amount. Thus, he selects the alternative that represents the maximum of the minima (or minimum of the maxima in case of loss) payoff in case of profits. The working method is summarized as follows:

(a) Locate the minimum (or maximum in case of profit) payoff value in case of loss (or cost) data corresponding to each alternative, then

(b) Select an alternative with the best anticipated payoff value (maximum for profit and minimum for loss or cost).

Since in this criterion the decision-maker is conservative about the future and always anticipates worst possible outcome (minimum for profit and maximum for cost or loss), it is called a *pessimistic decision criterion.* This criterion is also known as Wald's criterion.

Equally Likely Decision (Laplace) Criterion

Since the probabilities of states of nature are not known, it is assumed that all states of nature will occur with equal probability, *i.e.,* each state of nature is assigned an equal probability. As states of nature are mutually exclusive and collectively exhaustive, so the probability of each of these must be l/(number of states of nature). The working method is summarized as follows:

(a) Assign equal probability value to each state of nature by using the formula:

$$1 \div (\text{number of states of nature}).$$

(b) Compute the expected (or average) payoff for each alternative (course of action) by adding all the payoffs and dividing by the number of possible states of nature or by applying the formula:

(Probability of state of nature j) × (Payoff value for the combination of alternative, i and state of nature j.)

(c) Select the best expected payoff value (maximum for profit and minimum for cost).

This criterion is also known as the criterion of insufficient reason because, except in a few cases, some information of the likelihood of occurrence of states of nature is available.

Criterion of Realism (Hurwicz Criterion)

This criterion suggests that a rational decision-maker should be neither completely optimistic nor pessimistic and, therefore, must display a mixture of both. Hurwicz, who suggested this criterion, introduced the idea of a coefficient of optimism (denoted by α) to measure the decision-maker's degree of optimism. This coefficient lies between 0 and 1, where 0 represents a complete pessimistic attitude about the future and 1 a complete optimistic attitude about the future. Thus, if a is the coefficient of optimism, then (1 – a) will represent the coefficient of pessimism.

The Hurwicz approach suggests that the decision maker must select an alternative that maximizes

H (Criterion of realism) $= \alpha$ *(Maximum in column)* + $(1 - \alpha)$ *(Minimum in column)*

The working method is summarized as follows:

(a) Decide the coefficient of optimism α (alpha) and then coefficient of pessimism $(1 - \alpha)$.

(b) For each alternative select the largest and lowest payoff value and multiply these with α and $(1 - \alpha)$ values, respectively. Then calculate the weighted average, H by using above formula.

(c) Select an alternative with best anticipated weighted average payoff value.

Criterion of Regret (Savage Criterion)

This criterion is also known as opportunity loss decision criterion or minimax regret decision criterion because decision-maker feels regret after adopting a wrong course of action (or alternative) resulting in an opportunity loss of payoff. Thus, he always intends to minimize this regret. The working method is summarized as follows:

(a) From the given payoff matrix, develop an opportunity-loss (or regret) matrix as follows:

(i) Find the best payoff corresponding to each state of nature, and

(ii) Subtract all other entries (payoff values) in that row from this value.

(b) For each course of action (strategy or alternative) identify the worst or maximum regret value. Record this number in a new row.

(c) Select the course of action (alternative) with the smallest anticipated opportunity-loss valued.

POSTERIOR PROBABILITIES AND BAYESIAN ANALYSIS

The search and evaluation of decision alternatives often reveal new information. If such information is regarding the identification of alternatives, it requires revision and expansion of the test of alternatives. But if it is regarding the effects of alternatives, consequences are restated. When the uncontrollable factors are involved, either the states of nature themselves are reconsidered or their likelihoods are revised. The value of new information is evaluated in terms of its impact on the expected payoff. The expected value and the cost of the new information are compared to determine whether it is worth acquiring.

An initial probability statement to evaluate expected payoff is called a *prior probability distribution.* The one which has been revised in the light of new information is called a *posterior probability distribution.* It will be evident that what is a posterior to one sequence of state of nature becomes the prior to others which are yet to happen.

This section will be concerned with the method of computing posterior probabilities from prior probabilities using a mathematical formula called Baye's theorem. A further analysis of problems using these probabilities with respect to new expected payoffs with additional information is called *prior-posterior analysis.* The Baye's theorem in general terms can be stated as follows:

Let A_1, A_2,..., A_n be mutually exclusive and collectively exhaustive outcomes. Their probabilities $P(A_1)$, $P(A_2)$, ..., $P(A_n)$ are known. There is an experimental outcome B for which the conditional probabilities $P(B \mid A_1)$, $P(B \mid A_2)$,.. ., $P(B \mid A_n)$ are also known. Given the information that outcome B has occurred, the revised conditional probabilities of outcomes A_1, *i.e.,* $P(A_i \mid B)$, i = 1 + 2, ..., n are determined by using following conditional probability relationship:

$$P\,(A_i \mid B) = \frac{P(A_i \text{ and } B)}{P(B)} = \frac{P(A_i \cap B)}{P(B)}$$

where $P(B) = P(A_i \cap B) + P(A_2 \cap B) + ... + P(A_n \cap B)$.

Since each joint probability can be expressed as the product of a known marginal (prior) and conditional probability,

$$P(A_i \cap B) = P(A_i) \times P(B \mid A_i)$$

Thus $$P(A_i|B)=\frac{P(A_i)P(B|A_i)}{P(A_1)\,P(B|A_1)+P(A_2)P(B|A_2)+...+P(A_n)P(B|A_n)}$$

DECISION MAKING UNDER RISK

Decision making under risk is a probabilistic decision situation, in which more than one state of nature exists and the decision-maker has sufficient information to assign probability values to the likely occurrence of each of these states. Knowing the probability distribution of the states of nature, the best decision is to select that course of action which has the largest expected payoff value.

The most widely used criterion for evaluating various courses of action (alternatives) under risk is the *Expected Monetary Value (EMV) or Expected Utility.*

Expected Monetary Value (EMV)

The expected monetary value (EMV) for a given course of action is the weighted average payoff, which is the sum of the payoffs for each course of action multiplied by the probabilities associated with each state of nature. Mathematically EMV is stated as follows:

$$\text{EMV (Course of action, } S_j) = \sum_{i=1}^{m} p_{ij}\, p_i$$

where m = number of possible states of nature

pj = probability of occurrence of state of nature i

p_{ij} = payoff associated with state of nature N_i and course of action, S_j

Steps for calculating EMV : The various steps involved in the calculation of EMV are as follows:

(a) Construct a payoff matrix listing all possible courses of action and states of nature. Enter the conditional payoff values associated with each possible combination of course of action and state of nature along with the probabilities of the occurrence of each state of nature.

(b) Calculate the EMV for each course of action by multiplying the conditional payoffs by the associated probabilities and add these weighted values for each course of action.

(c) Select the course of action that yields the optimal EMV.

Expected Opportunity Loss (EOL)

An alternative approach to maximizing expected monetary value (EMV) is to minimize the expected opportunity loss (EOL), also called *expected value of regret*. The EOL is defined as the difference between the highest profit (or payoff) for a state of nature and the actual profit obtained for the particular course of action taken. In other words, EOL is the amount of payoff that is lost by not selecting the course of action that has the greatest payoff for the state of nature that actually occur. The course of action due to which EOL is minimum is recommended.

Since EOL is an alternative decision criterion for decision-making under risk, therefore, the results will always be the same as those obtained by EMV criterion discussed earlier. Thus, only one of the two methods should be applied to reach a decision. Mathematically, it is stated as follows:

$$\text{EOL } (\textit{State of nature}, N_i) = \sum_{i=1}^{m} l_{ij} p_i$$

where l_{ij} = opportunity loss due to state of nature, N_i and course of action, S_j

p_j = probability of occurrence of state of nature, N_i

Steps for calculating EOL : Various steps involved in the calculation of EOL are as follows:

(a) Prepare a conditional profit table for each course of action and state of nature combination along with the associated probabilities.

(b) For each state of nature calculate the conditional opportunity loss (COL) values by subtrapting each payoff from the maximum payoff for that outcome.

(c) Calculate EOL for each course of action by multiplying the probability of each state of nature with the COL value and then adding the values.

(d) Select a course of action for which the EOL value is minimum.

Expected Value of Perfect Information (EVPI)

In decision-making under risk each state of nature is associated with the probability of its occurrence. However, if the decision-maker can acquire *perfect (complete and accurate) information* about the occurrence of various states of nature, then he will be able to select a course of action that yields the desired payoff for whatever state of nature that actually occurs.

We have seen that the EMV or EOL criterion helps the decision-maker select a particular course of action that optimizes the expected payoff without any additional informa .ı. *Expected value of perfect information* (EVPI) represents the maximum amount of money the decision-maker has to pay to get this additional information about the occurrence of various states of nature before a decision has to be made. Mathematically it is stated as

EVPI = Expected profit (or value) with perfect information under certainty

– Expected profit without perfect information

$$\sum_{i=1}^{m} p_{ij} (N_i)\, p_i - EMV^*$$

where p_{ij} = best payoff for state of nature, N_i

p_i = probability of state of nature, N_i

EMV^* = maximum expected monetary value

DECISION TREE ANALYSIS

Decision-making problems discussed so far are referred to as single stage decision problems, because the payoffs, states of nature, courses of action and probabilities associated with the occurrence of states of nature are not subject to change.

However, situations may arise when a decision-maker needs to revise his previous decisions on getting new information and make a sequence of other decisions. Thus, the problem becomes a multi-stage decision problem because the consequence of one decision affects future decisions. For example, in the process of marketing a new product, the first decision is often test marketing and the alternative courses of action might be either intensive testing or gradual testing. Given various possible consequences—good, fair, or poor, the decision-maker may be required to decide between redesigning the product, an aggressive advertising campaign or complete withdrawal of product, etc. Given that decision, there will be an outcome which leads to another decision and so on.

Decision tree analysis involves construction of a diagram showing all the possible courses of action, states of nature and the probabilities associated with the states of nature. The *decision diagram* looks very much like a drawing of a tree, therefore also called *decision-tree*.

A decision tree consists of *nodes, branches, probability estimates*, and *payoffs*. There are two types of nodes: *decision nodes* and chance nodes. A decision node is usually represented by a square and indicates places where a decision-maker must make a decision. Each branch leading away from a decision node represents one of the several possible courses of action available to the decision-maker. The chance node is represented by a circle and indicates a point at which the decision-maker will discover the response to his decision, *i.e.*, different possible outcomes occurred from a chosen course of action.

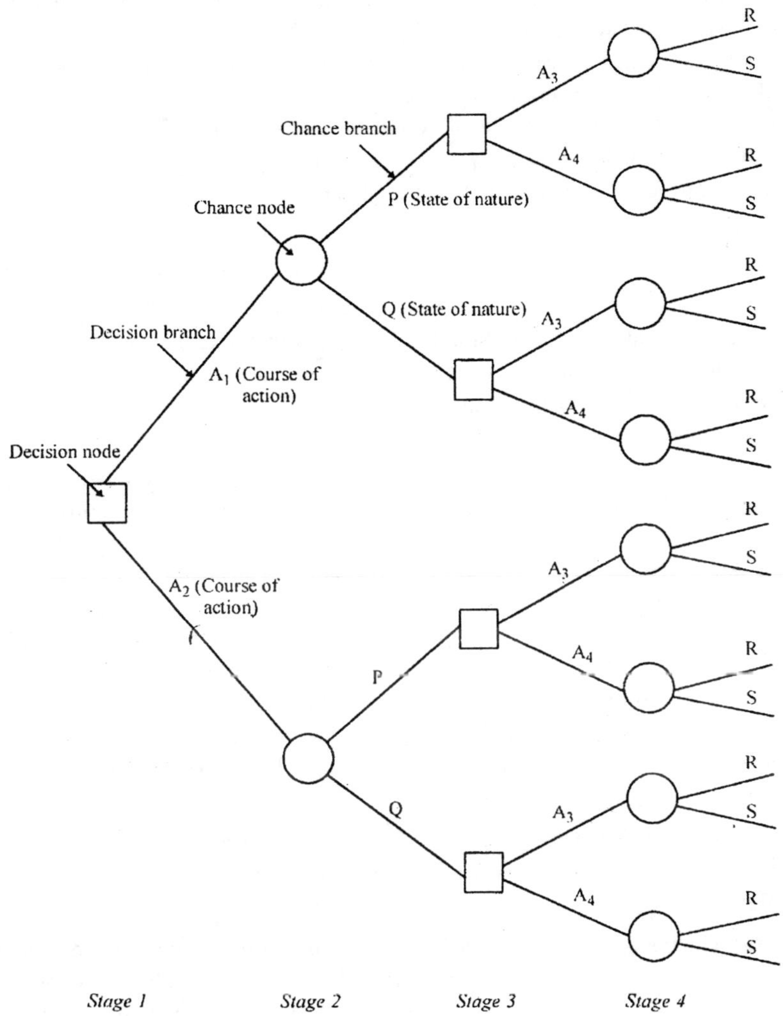

Fig. 5.1 : Decision Tree.

Branches emanate from and connect various nodes and are either decisions or states of nature. There are two types of branches: *decision branches* and *chance branches*. Each branch leading away from a decision node represents a course of action or strategy that can be chosen at a decision point, whereas a branch leading away from a chance node represents the state of nature of a set of chance factors. Associated probabilities are indicated along side of respective chance branch. These probabilities are the likelihood that the chance outcome will assume the value assigned to the particular branch. Any branch that makes the end of the decision tree, *i.e.*, it is not followed by either a decision or chance node, is called a *terminal branch.* A terminal branch can represent either a course of action or a chance outcome. The terminal points of a decision tree are said to be mutually exclusive points so that exactly one course of action will be chosen.

The *payoff* can be positive (*i.e.*, revenue or sales) or negative (*i.e.*, expenditure or cost) and they can be associated either with decision or chance branches.

An illustration of a decision tree is shown in Fig. 5.1. It is possible for a decision tree to be deterministic or probabilistic and it can further be divided into single stage (a decision under condition of certainty) and a multistage (a sequence of decisions).

The optimal sequence of decisions in a tree is found by starting at the right-hand side and rolling backward. The aim of this operation is to maximize the return from the decision situation. At each node, an expected return is calculated (called *position value*). If the node is a chance node, then the position value is calculated as the sum of the products of the probabilities or the branches emanating from the chance node and their respective position values. If the node is a decision node, then the expected return is calculated for each of its branches and the highest return is selected. The procedure continues until the initial node is reached. The position values for this node corresponds to the maximum expected return obtainable from the decision sequence.

Remark : Decision trees versus probability trees. Decision trees are basically an extension of probability trees. However, there are several basic differences:

(i) The decision tree utilizes the concept of 'rollback' to solve a problem. This means starting at the right-hand terminus with the highest expected value of the tree and working back to the current or beginning decision point to determine the decision or decisions that should be made. Most decisions require trees with numerous branches and more than one decision point. It is the multiplicity of decision points that make the rollback process necessary.

(ii) The probability tree is primarily concerned with calculating the correct probabilities, whereas the decision tree utilizes probability factors as a means of arriving at a final answer.

(iii) A most important feature of the decision tree, not found in probability trees, is that it takes time differences of future earnings into account. At any stage of the decision tree, it may be necessary to weigh differences in immediate cost or revenue against differences in value at the next stage).

DECISION MAKING WITH UTILITIES

So far problems were analyzed with probabilistic states of nature where selection of an optimal course of action was based on the criterion of expected profit (or loss) expressed in monetary terms. However, in many situations, such criteria involving expected monetary payoff may not be appropriate. This is because of the fact that different individuals attach different utility to money under different conditions. The term utility is *the measure of preference for various alternatives in terms of relative value for money*. The utility of a given alternative is unique to individual decision-makers and unlike a simple monetary amount, can incorporate intangible factors or subjective standards from their own value systems.

Illustration : Mr. Ram has won Rs. 1,000 in a quiz programme. In the last he is asked either to compete or quit. Now he has two alternatives: (a) quit and take his winnings, (b) take a last chance in which he has 50-50 chance of winning Rs. 4,000 or nothing, then the question is: What would he do? On EMV basis he has

$$\text{EMV (A)} = 0.50\ (4{,}000) + 0.50\ (0) = \text{Rs } 2{,}000$$

This amount is twice what he has already won. But would he really give up Rs 1,000 for a 50-50 chance or Rs. 4,000 or nothing? Many individuals would not because they would think of all the alternatives they could do with Rs 1,000 and how they would regret it if they end up with nothing. Hence a new payoff measure utility reflecting the decision-maker's attitude and preference has to be introduced.

The basic axioms of utility may be stated as follows:

(i) IT outcome A is preferred to outcome B, then the utility U(A) of outcome A is greater than the utility U(B) of outcome B, and vice-versa. If both are equally preferred then

$$U(A) = U(E)$$

(ii) If the decision-maker is indifferent between the two alternatives and outcome A is received with probability p, and outcome C with probability (1 – p), then

$$U(B) = pU(A) + (1 - p)U(C)$$

Under this alternative criterion, it is assumed that a rational decision-maker will choose that alternative which optimizes the expected *utility* rather than expected monetary value. Once we know the individual's utility function along with the probability assigned to outcomes in a particular situation, then total expected utility for each course of action can be obtained by multiplying utility values with their probabilities. The strategy which corresponds to optimum utility function is called the *equal strategy*.

Utility Functions

Utility function is a formula or method to describe the relative preference value that individuals have for a given criterion such as money, goods, etc. Preferences are often determined by proposing a situation whereby decision-maker must choose between receiving a given amount, say Rs. 20,000 for a certain thing versus a 50-50 chance of gaining a larger amount or nothing, say Rs. 60,000 or zero. The gamble amount of Rs. 60,000 is then adjusted upward or downward until the individual is indifferent to whether decision-maker receives the certain amount of Rs. 20,000 or the gamble. This indifference point then establishes one experimentally determined value on the individual's utility curve and other points are similarly determined.

Once derived, a utility function can be used to convert a decision criteria value into utilise so that a decision can be made on the basis of maximizing the expected utility value (EUV) rather than, say EMV.

Construction of Utility Curves

Suppose, there is a project which will payoff with Rs 50,000 or- Rs 50,000. Then what probability of success would make the decision-maker just indifferent about whether to undertake the project or not? Suppose he says that he would require an 80 per cent probability of success to make him indifferent about undertaking the project. Thus, we need to associate a utility value to one payoff. Let the utility of Rs 5,000 be 10, then it follows that

$$U(\text{Re } 0) = (0.80)\, U\,(\text{Rs } 50{,}000) + (0.20)\, U\,(-\text{Rs } 50{,}000)$$

or $\quad 0 = 0.80\,(10) + (0.20)\, U\,(-\text{Rs } 50{,}000)$

or $\quad U\,(-\text{Rs } 50{,}000) = -40$

Now we have three points on the utility curve

$$U\,(+50{,}000) = 100;\ U(0) = 0 \text{ and } U\,(-50{,}000) = -40.$$

Other points on the curve can be obtained by asking such questions to the decision-maker. The procedure of associating utility value to a payoff can be made more realistic by attempting to relate it to either past or current projects.

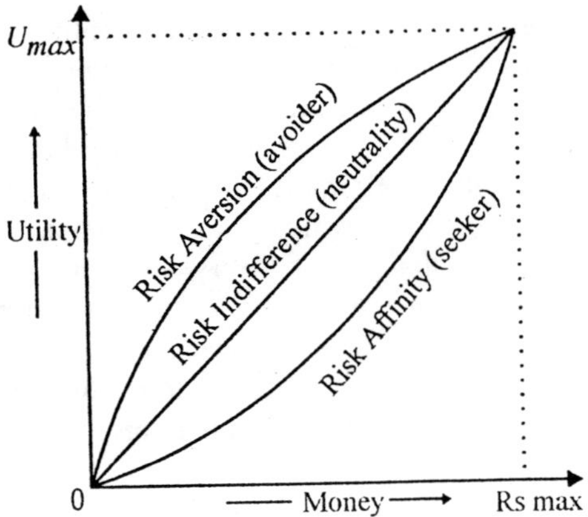

Fig. 5.2 : Utility Curve and Associated Risk Preferences.

Utility Curve

Suppose, a loss of Rs. 90,000 causes a businessman to be declared bankrupt. If he was able to withstand fairly (or comfortably) the loss of Rs. 9,000, then that loss of Rs. 90,000 is ten times worse than the loss of Rs. 9,000. In other words, the concept of utility must be developed in such a way that conversion of monetary units into utils reflects such preferences.

A utility curve that relates utility values to rupee values is constructed. Such a curve usually is obtained by placing the decision-maker in various hypothetical decision situations and, plotting the decision-maker's pattern of choices in terms of risk and utilities. Figure 5.7 shows several utility curves and the risk preference associated with each.

Suppose, the relationship between monetary gains, losses and utilities for gains and for small negative losses is established. As shown in Figure 5.7, if the curve is bent down non-linearly, then we assign to large losses a disproportionately large negative utility. It is important not to make the curve bend down too steeply or to start the bending too quickly, since this could lead individuals into a situation where they attach such a heavy weighting to the possibility of loss that they never take any risk and thus, never make any gains.

On the positive side of the curve, it is usual for the curve to eventually bend away from the straight line. This indicates that increasing units of money are resulting in smaller additional gains in utility.

GAMES

The term 'game' refers to the general situation of conflict and competition in which two or more competitors (or participants) are involved in decision-making activities in anticipation of certain outcomes over a period of time. The competitors are referred as *players*. A player may be an *individual*, a *group of individuals*, or an *organization*. Few examples of competitive and conflicting decision environment involving the interaction between two or more competitors where techniques of theory of games may be used to resolve them are: (i) pricing of products, where a firm's ultimate sales are determined not only by the price levels it selects but also by the prices its competitors set, (ii) various TV networks have found that program success is largely dependent on what the competitors presents in the same time slot; the outcomes of one networks programming decisions have, therefore, been increasingly influenced by the corresponding decisions made by other networks, (iii) success of a business tax strategy depends greatly on the position taken by the internal revenue service regarding the expenses that may be disallowed, (iv) success of an advertising/marketing campaign depends largely on various types of services offered to the customers, etc.

The area of academic study, *theory of games* provides a series of mathematical models that may be quite useful in explaining interactive decision-making concepts. But as a practical tool, it is limited in scope. However, such models provide an opportunity to a competitor to evaluate not only his personal alternatives (courses of action), but also the evaluation of the opponent's (or competitor's) possible choices in order to win the game. The models in the *games theory* can be classified depending upon the following factors:

Strategy : Strategy for a player is the list of all possible actions (moves or courses of action) that he will take for every payoff (outcome) that might arise. It is assumed that the rules governing the choices are known in advance to the players. The outcome resulting from a particular choice is also known to the players in advance and is expressed in terms of numerical values (*e.g.*, money, per cent of market share or utility). Here it is not necessary that players have definite information about each other's strategies.

Sum of gains : The game sum of gains to one player is exactly equal to the sum of losses to another player, so that sum of the gains and losses equals zero, then the game is said to be a *zero-sum gone*. Otherwise it is said to be non-zero sum game.

Number of players : If a game involves only two players (competitors), then it is called a *two-person game*. However, if the number of players is more, the game is referred to as n-person game.

The particular strategy (or complete plan) by which a player optimizes his gains or losses without knowing the competitor's strategies is called *optimal strategy*. The expected outcome per play when players follow their optimal strategy is called the *value of the game*.

Generally, two types of strategies are employed by players in a game.

(b) *Mixed Strategy* : Courses of action that are to be selected on a particular occasion with some fixed probability are called *mixed strategies*. Thus, there is a probabilistic situation and objective of the players is to maximize expected gains or to minimize expected losses by making choice among pure strategies with fixed probabilities.

(a) *Pure Strategy* : It is the decision rule which is always used by the player to select the particular strategy (course of action). Thus, each player knows in advance of all strategies out of which he always selects only saddle point one particular strategy regardless of the other player's strategy, and the objective of the players is to maximize gains or minimize losses.

INVESTMENT ANALYSIS

Efficient allocation of funds is one of the most important functions of the financial management in modem times. This function involves the firm's decision to invest its funds in long-term assets and other profitable activities. Such investment decisions are of great significance for the firm as they affect its wealth, influence its size, set the pace and direction of its growth and determine its business risk. *Thus investment decisions involve the most efficient investment of the funds in long-term activities in anticipation of the expected flow of future benefits over a number of years*. The future benefits are measured in terms of cash flows. In the investment decisions, it is the flow of cash—outflow and inflow—and not the accrued earnings which is important

Investment decisions influence the firm's wealth. If the investment proposals are profitable, the firm's wealth will increase, otherwise it will decrease. The investment decisions, thus, affect the value of the firm. Every firm would therefore like to undertake *investment* analysis which involves the consideration of investment proposals, estimation of cash flows for the proposals, evaluation of cash flows, selection of projects based on some criterion and finally the continuous revaluation of these projects.

METHODS OF INVESTMENT ANALYSIS

The most widely used methods of evaluating an investment proposal

A. Deterministic Models

1. Break-Even Analysis
2. Pay-back Period Methods
3. Average Rate of Return Method
4. Discounted Cash Flow Methods
   (a) Net Resent Value Methods
   (b) Internal Rate of Return Method
5. Discounted Payback Period Method

B. Probabilistic Models

1. Risk Adjusted Discount Kale Methods
2. Certainty-Equivalent Approach
3. Expected Monetary Value Methods
4. Butler's and Hertz's Models.

BREAK-EVEN ANALYSIS

The break-even analysis is the most widely known form of cost-volume-profit (CVP) analysis. That is why the two terms are used interchangeably by many. The break-even analysis provides a relationship between revenues and costs with respect to volume (quantity) of sales. It represents the level of sales at which costs and revenues are in equilibrium; the equilibrium point being known as the break-even point (BEP). At the break-even point, total revenue is equal to the total costs; it is a no-profit, no-loss point. This analysis assumes that the total costs can be separated as fixed costs and variable costs.

Two approaches can be used to determine the break-even point:

1. The formula approach
2. The chart approach.

1. *The Formula Approach* : The break-even point can be computed in terms of units or intends of money value of sales volume or as percentage of can mated capacity.

   (i) *In units* : For a single-product firm, the break-even point in terms of units will be reached when the total earned revenue becomes equal to Ac total costs. Ifs denotes the unit selling price, v the unit variable cost, F the fixed costs and Q_B the breack-even point (units), then

   total revenue = $Q_B.s$, and

   total revenue = $F + Q_B.v$

Therefore, at break-even point, $Q_B.s = F + Q_B.v$

or $$Q_B = \frac{F}{s - v}$$

or breack-even point (units) $= \dfrac{\text{Fixed costs}}{\text{selling price/unit - variable cost/unit}}$

(ii) *In Money Value :* The break-even point for a single-product firm can also be expressed in terms of rupee value of sales volume. If both sides of multiplied by the unit selling price, we get the break-even point in terms of rupees. Thus

$$R_B = Q_B.s = \frac{F}{s - v}.s$$

As both the variable costs and sales revenues vary in direct proportion to sales volume, the above formula can also be used for a multi product firm. For such a firm,

break-even point (rupees) $= \dfrac{\text{Fixed costs}}{\text{1 - Total variable cost/Total sales revenue}}$

(iii) *As a Percentage of Capacity :* The break-even point as a percentage of capacity is obtained by dividing the break-even sales by the total capacity sales. Thus

break-even point, (as a percentage of capacity) $= \dfrac{\text{Break - even sales}}{\text{Capacity sales}}$

2. *The Chart Approach :* The break-even chart represents a pictorial view of the relationship between costs, volume and profit. The break-even point in this chart is the one at which the total cost line and the total sales line intersect. The break-even chart is constructed as follows:

   (i) Represent the sales revenue along the horizontal axis. The sales revenue may be expressed in terms of units, rupees or as a percentage of capacity.

   (ii) Represent the revenue, faced and variable costs along the vertical axis. A similar vertical line may be drawn on the right of the chart to complete the square.

   (iii) Draw the fixed cost line parallel to the horizontal axis through the fixed cost point.

   (iv) Draw the total sates line and the total cost line. The point at which they intersect is the break-even point. The angle between these lines a called the angle of incidence. Larger this angle, lower .the break-even point and vice-versa. The area to the left of this point is the *loss area* and represents the unrecovered fixed costs, while the area to its right is the *profit area.*

The excess actual or budgeted sales over the break-even sales is called *margin of safety*. The margin of safety ratio is expressed as

$$\text{M/s ratio} = \frac{\text{Budgeted sales - B/E sales}}{\text{Budgeted sales}}$$

The margin of safety indicates the extent to which sales may fall before a firm suffers a loss. Larger the margin of safety, safer the firm.

ASSUMPTIONS IN BREAK-EVEN ANALYSIS

The above discussion of the break-even analysis is based on the following assumption.

1. The total cost can be separated into fixed and variable costs.
2. The total fixed cost remains unchanged with changes in sales volumes.
3. The variable cost per unit is constant and the total variable cost is proportional to the sales volume.
4. The selling price per unit is constant *i.e.,* it does not change with volume.
5. The firm manufactures only one product or if there are multiple products, the sales mix does not change.

Utility of Break-Even Analysis

Break-even analysis is the most useful technique of profit planning and control. It is a device to explain the relationship between cost, volume and profits. The Utility of the break-even analysis lies in the following advantages it has:

1. It is simple device to understand complicated accounting data.
2. It is a simple diagnostic tool. It indicates to management the causes of increasing break-even point and falling profits. An analysis of these causes can reveal to management what actions should be taken.
3. It provides basic information for further profit improvement studies and is a useful starting point for detailed investigations.
4. It is a useful method for considering the risk implications of alternative actions. From one alternative a firm may expect a higher profit and, also a higher break-even point, while another alternative may produce comparatively lower profit but may also entail a lower break-even point. In taking a decision, the firm should not only consider the profits expected from the alternative but also the probability of reaching the break-even point. An alternative for which the *probability* of reaching the break-even point is higher should be preferred.

Limitations of Break-Even Analysis

1. It is difficult to separate costs into fixed and variable components.
2. The total fixed cost may not remain unchanged over the entire volume range.
3. The unit selling price and unit variable cost may not remain constant.
4. It is difficult to use for a multi-product firm.
5. It is a short-term concept and has limited use in long-term planning.
6. It is a static tool and shows the relationship between costs volume and profits of a firm at given point of time only.

PAYBACK PERIOD METHOD

Payback period is one of the most popular traditional method of evaluating investment proposals. *It can be defined as the number of years required to recover the cash invested in a project.* If the annual cash inflows are same, the payback period can be computed dividing cash invested by the annual cash inflow; if unequal, it can be calculated by adding up the cash inflows until the total is equal to the initial cash invested.

Payback period can be used as an accept-or-reject rule and also as a method of ranking projects. If the payback period, calculated for a project is less than the maximum payback period setup by the management, the project is accepted; if not, it is rejected. As a ranking method it gives highest ranking to the project that has the shortest payback period and lowest ranking to the project with highest payback period.

Merits

1. It is simple to understand and easy to calculate,
2. It is less costly than the other evaluation techniques.
3. It also makes clear that no project yields profit unless the payback periods over. It thus emphasises that projects in which cash inflow accrue quickly arc profitable. It is quite useful for projects where the risk of obsolescence is high, *e.g.*, computers and aircraft manufacturing.

Demerits

1. It takes no account of the cash inflows that accrue after the payback period and is not an appropriate method of measuring the probability of an investment project.
2. It fails to consider the magnitude and tuning of cash inflows *i.e.*, it gives equal weight to returns of equal amounts even though they take place in different periods.

AVERAGE (ACCOUNTING) RATE OF RETURN METHOD

The accounting rate of return (ARR) is found by dividing the average annual net income (after taxes and depreciation) by the average cash outlay *i.e.*, average book value after depreciation. The accounting rate of return is thus an average rate and can be determined by the following equation.

$$\text{ARR} = \frac{\text{Average income}}{\text{Average investment}}$$

A variation of the ARR method is to divide the average earnings after taxes by the original cost of the project instead of average cost.

This method accepts all those projects whose ARR value is higher than the minimum rate established by the management and rejects those whose ARR value is lower. It also ranks the project as number one if it has highest. ARR and lowest rank would be assigned to the project with lowest ARR.

Merits

1. It is very simple to understand and use.
2. It can be easily calculated using the accounting data.
3. It uses the entire stream of incomes in calculating the accounting rate.

Demerits

1. It uses accounting profits and not cash flows in comparing the projects.
2. It ignores the time value of money. Profits accruing in different periods are valued equally.
3. It does not allow for the fact that the profits can be recovered.

TIME-ADJUSTED OR DISCOUNTED CASH FLOW (DCF) METHODS

The above methods suffer from a serious drawback—they fail to recognise the time value of money in evaluating the investment worth of the projects. Methods that fully recognise the value of money are called *time-recognised* or *discounted cash flow* or *present value methods*. They are

1. Net present value method.
2. Internal rate of return method.

Net Present Value (NPV) Method

It is one of the discounted cash flow methods and fully recognises the time value of money. It correctly postulates that cash flows arising at different time periods differ in values and are comparable only when their equivalents—present values—are found out It involves the following steps:

1. Select an appropriate rate of interest to discount cash flows. Generally, this rate is the firm's *cost of capital* which equals the minimum rate of return expected by the investor.
2. Compute present value of cash inflows and outflows using the cost of capital as the discount rate. If all cash outflows are made in the initial year, then their present value will be equal to the cash actually spent.
3. Find out the net present value by subtracting the present value of cash outflows from the present value of cash inflows.
4. Accept the in vestment project its net present value is positive or zero, and reject it if its net present value is negative.

The equation for NPV, assuming that all cash outflows are made in the initial year t_0 can be written as

$$NPV = \left[\frac{A_1}{1+k} + \frac{A_2}{(1+k)^2} + \ldots + \frac{A_n}{(1-k)^n}\right] - C$$

$$= \sum_{t=1}^{n} \frac{A}{(1+k)^t} - C_1$$

where A_1, A_2, ... represent the cash inflows, k is the firm's cast of capital, C is the cash outlay and n is the expected life of the proposal

Merits

1. It recognises the time value of money.
2. It considers all cash flows over the entire life of the project.
3. It aims at maximizing the welfare of the owners of the organisation.

Demerits

1. It is difficult to use.
2. It assumes that the discount rate-which is usually the firm's cost of capital—is known. But the cost of capital is difficult to understand and measure.
3. It may not give satisfactory answer when projects involving different amount of investment are compared. The project with higher NPV may not be desirable if it also requires a large investment.

INTERNAL RATE OF RETURN (IRR) METHOD

The internal rate of return is the rate that equals the present value of the cash in rows with the present value of cash outflows of an investment.

Thus, it is the rate at which the net present value of the investment is zero. It can be determined from the following equation:

$$C = \frac{A_1}{1+r} + \frac{A_2}{(1+)^2} + \ldots + \frac{A_n}{(1+r)^n} = \sum_{t=1}^{n} \frac{A_t}{(1+r)^r}$$

or
$$\sum_{t=1}^{n} \frac{A_t}{(1+r)^r} - C = 0$$

It may be seen that the IRR formula is the same as used for the NPV method with the difference that in the NPV method the required rate of interest (cost of capital), k is known and the NPV is to he found out, while in the IRR method the value of r is to be determined at which the NPV is zero. The value of r in the IRR equation is determined by trial and error method. Some value of the rate of return is chosen and NPV of cash inflows is calculated. If this value is lower than the present value of cash outflows, a lower rate is tried; if this value is higher than the present value of cash outflows, a higher rate is tried. The process is repeated until the net present value becomes zero.

According to IRR method, a project is accepted if its internal rate of return is higher than or equal to the minimum required rate of return (*i.e.*, $r \geq k$); and the project is to be rejected if $r < k$.

Merits

1. Like the NPV method, it considers the time value of money.
2. It considers cash flows over the entire life of the project.
3. It has psychological appeal to true users.

Demerits

1. It is difficult to understand and use in practice.
2. It may yield results inconsistent with the NPV method if the project differ in their expected lives or cash outlays or timing off.
3. It may not give unique answers in all situations.

SOLVED EXAMPLES

Example 1:

A manufacturing firm produces a single product whose selling price is Rs. 16 per unit and the variable costs per unit are Rs. 12. If the annual fixed costs of the firm are estimated as Rs. 1,20,000, find the break-even point in units, in rupees and as a percentage of capacity if the firm has an estimated capacity of 50.000 units of the product. What is the margin of safety?

Solution:

Here, s = Rs. 16, v = Rs. 12 F = Rs. 1,20,000.

$\therefore$ Break-even point (units), $Q_s = \frac{F}{s-v} = \frac{1,20,000}{16-12} = 30,000$ units.

Break-even point (rupees), $R_B = \frac{F}{1-\frac{v}{s}} = \frac{1,20,000}{1-\frac{12}{16}} =$ Rs. 4,80,000.

Break-even point (as a percentage of capacity) $= \frac{\text{Break - even sales}}{\text{Capacity sales}}$

$$= \frac{30,000}{50,000} = 60\%$$

Margin of safety = 50,000 – 30,000 = 20,000 units.

Example 2:

A firm manufactures three types of products. The fixed and variable costs are given below:

		Fixed Cost (Rs.)	***Variable Cost per Unit (Rs.)***
Product A	:	*25,000*	*12*
Product B	:	*35,000*	*9*
Product C	:	*53,000*	*7*

The likely demand (units) of the products is given below:

Poor demand	:	*3,000*
Moderate demand	:	*7,000*
High demand	:	*11,000*

If the sale price of each type of product is Rs. 25, then prepare the payoff matrix.

Solution:

Let D_1, D_2, and D_3 be the poor, moderate and high demand, respectively. Then payoff is given by

Payoff = Sales revenue – Cost

The calculations for payoff (in thousand) for each pair of alternative demand (course of action) and the types of product (state of nature) are shown below:

$D_1A = 3 \cdot 25 - 25 - 3 \times 12 = 14$ $\quad$ $D_2A = 7 \times 25 - 25 - 7 \times 12 = 66$

$D_1B = 3 \times 25 - 35 \quad \times 9 = 13$ $D_2B = 7 \times 25 - 35 - 7 \times 9 = 77$

$D_1C = 3 \times 25 - 53 - 3 \times 7 = 1$ $D_2C = 7 \times 25 - 53 - 7 \times 7 = 73$

$D_3A = 11 \times 25 - 25 - 11 \times 12 = 118$

$D_3B = 11 \times 25 - 35 - 11 \times 9 = 141$

$D_3C = 11 \times 25 - 53 - 11 \times 7 = 145$

The payoff values are shown in Table 5.2.

Table 5.2 ('000 Rs.)

Product Type	*Alternative Demand*		
	D_1	D_2	D_3
A	14	66	118
B	13	77	141
C	1	73	145

Example 3:

Suppose a firm has the following budget for a particular year.

Sales (1.00,000 units @ Rs. 20) = Rs. 20,00,000,

variable costs (1,00,000 units @ Rs. 10) = Rs. 10,00,000.

fixed costs = Rs. 4,00.000.

net profits = Rs. 6.00.000.

What shall be the impact on the firm's profits, if the following changes lake place:

Increase in price: 20%,

decrease in sales volume: 25%,

increase in variable costs: 10%,

increase in fixed costs: 5%.

Solution:

Selling price per unit, s = Rs. (20 × 1.2) = Rs. 24.

Sales volume, n = 1,00,000 × 0.75 = 75,000 units.

Variable cost/unit, v = Rs. (10 × 1.1) = Rs. 11.

Fixed costs, F = Rs. (4,00.000 × 1.05} = Rs. 4,20,000.

∴ Profit (s – v)n n – F = Rs. 1(24 – 11) × 75.000 – 4.20.000]

= Rs. 5,55,000

$\therefore$ Percent change in profit $= \dfrac{5,55,000 - 6,00,000}{6,00,000} \times 100 = -7.5\%$

Example 4:

A food products company is contemplating the introduction of a revolutionary new product with new packaging or replace the existing product at much higher price (S_1) or a moderate change in the composition of the existing product with a new packaging at a small increase in price (S_2) or a small change in the composition of the existing product except the word 'New' with a negligible increase in price (S_3). The three possible states of nature or events are: (i) high increase in sales (N_1), (ii) no change in sales (N_2) and (iii) decrease in sales (N_3). The marketing department of the company worked out the payoffs in terms of yearly net profits for each of the strategies of three events (expected sales). This is represented in the following table:

	States of Nature		
Strategies	N_1	N_2	N_3
S_1	*7,00,000*	*3,00,000*	*1,50,000*
S_2	*5,00,000*	*4,50,000*	*0*
S_3	*3,00,000*	*3,00,000*	*3,00,000*

Which strategy should the concerned executive choose on the basis of

(i) *Maximin criterion*

(ii) *Maximax criterion*

(iii) *Minimax regret criterion*

(iv) *Laplace criterion?*

Solution:

The payoff matrix is rewritten as follows:

(i) Maximin Criterion

States of Nature		***Strategies***	
	S_1	S_2	S_3
N_1	7,00,000	5,00,000	3,00,000
N_2	3,00,000	4,50,000	3,00,000
N_3	1,50,000	0	3,00,000
Column minimum	1,50,000	0	**3,00,000** ← Maximin

The maximum of column minima is 3,00,000. Hence, the company should adopt strategy S_3

(ii) Maximax Criterion

States of Nature	*Strategies* S_1	S_2	S_3
N_1	7.00.000	5.00.000	3,00,000
N_2	3.00.000	4.50.000	3,00,000
N_3	1.50.000	0	3,00,000
Column maximum	**7,00,000**	5.00.000	3,00,000
	↑ Maximax		

The maximum of column maxima is 7,00,000. Hence, the company should adopt strategy S_1.

(iii) *Minimax Regret Criterion* Opportunity loss table is shown below:

States of Nature	*Strategies* S_1	S_2	S_3
N_1	7,00,000 – 7,00,000 = 0	7.00,000 – 5.00,000 = 2,00,000	7,00,000 – 3,00,000 = 4,00,000
N_2	4,50,000 – 3,00,000 = 1,50,000	4,50,000 – 4,50,000 = 0	4,50,000 – 3,00,000 = 1,50,000
N_3	3,00,000 – 1,50,000 = 1,50,000	3,00,000 – 0 = 3,00,000	3,00,000 – 3,00,000 = 0
Column maximum	1,50,000	3.00,000	4,00,000
	↑ Minimax regret		

Hence the company should adopt minimum opportunity loss strategy, S_1.

(iv) Laplace Criterion Since we do not know the probabilities of states of nature, assume that they are equal. For this example, we would assume that each state of nature has a probability 1/3 of occurrence.

Thus,

Strategy	*Expected Return (Rs.)*	
S_1	(7,00,000 + 3,00,000 + 1.50,000)/3	= 3,83,333.33
S_2	(5.00,000 + 4,50,000 + 0)/3	= 3,16,666.66
S_3	(3.00,000 + 3.00,000 + 3.00,000)/3	= 3.00,000

Since the largest expected return is from strategy S_2 the executive must select strategy S_1.

Example 5:

A manufacturer makes a product, of which the principal ingredient is a chemical X. At the moment, the manufacturer spends Rs. 1,000 per year on supply of X, but there is a possibility that the price may soon increase to four times its present figure because of a worldwide shortage of the chemical. There is another chemical Y, which the manufacturer could use in conjunction with a third chemical, Z in order to give the same effect as chemical X. Chemicals Y and Z would together cost the manufacturer Rs. 3,000 per year, but their prices are unlikely to rise. What action should the manufacturer take? Apply the maximin and minimax criteria for decision-making and give two sets of solutions. If the coefficient of optimism is 0.4, find the course of action that minimizes the cost.

Solution:

The data of the problem is summarized in the following table (negative figures in the table represents profit).

States of Nature	***Courses of Action***	
	S_1 (use Y and Z)	S_2 (use X)
N_1 (Price of X increases)	–3,000	–4,000
N_2 (Price of X does not increase)	–3,000	–1,000

(i) Maximin Criterion

States of Nature	***Courses of Action***	
	S_1	***S_2***
N_1	–3,000	–4,000
N_2	3,000	–1,000
Column minimum	–3,000 ↑ Maximin	

Maximum of column minima = –3,000. Hence, the manufacturer should adopt action S_1.

(ii) Minimax (or opportunity loss) Criterion

States of Nature	***Courses of Action***	
	S_1	***S_2***
N_1	–3,000 – (–3,000) = 0	–3,000 – (–4,000) = 1,000
	– 1,000 – (–3,000) = 2,000	–1,00 – –1,000) = 0
Maximum opportunity	2,00	**1,000** ← Minimax

Hence, manufacturer should adopt minimum opportunity loss course of action S_2.

(iii) *Hurwicz Criterion* : Given the coefficient of optimism equal to 0.4, the coefficient of pessimism will be 1 – 0.4 = 0.6. Then according to Hurwicz, select course of action that optimizes (maximum for profit and minimum for cost) the payoff value

H = α (Best payoff) + (1 – α) (Worst payoff)

= α (Maximum in column) + (1 – α) (Minimum in column)

Course of Action	*Best Payoff*	*Worst Payoff*	*H*
S_1	–3,000	–3,000	–3.000
S_2	–1,000	–4,000	–2.800

Since course of action S_2 has the least cost (maximum profit) = 0.4(1,000) + 0.6(4,000) = Rs. 2,800, the manufacturer should adopt it.

Example 6:

Solve the game whose payoff matrix is given below:

		Player B		
Player A	B_1	B_2	B_3	B_4
A_1	3	2	4	0
A_2	3	4	2	4
A_3	4	2	4	0
A_4	0	4	0	8

Solution:

It is clear that this game has no saddle point. Therefore, we try to reduce the size of the given payoff matrix by using dominance principles.

From player A's point of view, first row is dominated by the third row yielding the reduced 3 × 4 payoff matrix. In the reduced matrix from player B's point of view, first column is dominated by the third column. Thus, by deleting the first row and then the first column, the reduced payoff matrix so obtained is:

		Player B	
Player A	B_2	B_3	B_4
A_1	4	2	4
A_3	2	4	0
A_4	4	0	8

Now it may be noted that none of the pure strategies of players A and B is inferior to any of their other strategies. However, the average of payoffs due to strategies B_3 and B_4, {(2 + 4)/2; (4 + 0)/2; (0 + 8)/2} = (3, 2, 4) is superior to the payoff due to strategy B_2 of player B. Thus, strategy B^ may be deleted from the matrix. The new matrix so obtained is:

	Player B	
Player A	B_3	B_4
A_2	2	4
A_3	4	0
A_4	0	8

Again in the reduced matrix, the average of the payoffs due to strategies A_3 and A_4 of player A, *i.e.*, $\{(4 + 0)/2; (0 + 8)/2\} = (2, 4)$ is the same as the payoff due to strategy A_2. Therefore, the player A will gain the same amount even if the strategy A_2 is never used. Hence, after deleting the strategy A_2 from the reduced matrix, a new reduced 2×2 payoff is obtained,

	Player B	
Player A	B_3	B_4
A_3	4	0
A_4	0	8

This game has -no saddle point. Let player A chooses his strategies A_3 and A_4 with probability p_1 and p_2, respectively such that $p_1 + p_2 = 1$. Also Jet player B choose his strategies with probability q_1 and q_2, respectively such that $q_1 + q_2 = 1$. Since both players want to retain their interests unchanged, therefore, we may write:

$4p_1 + 0.p_2 = 0.p_1 + 8p_2$ $\qquad$ $4q_1 + 0.q_2 = 0.q_1 + 8q_2$

$4p_1 = 8(1 - p_1)$ $\qquad$ $4q_1 = 8(1 - q_1)$

$p_1 = 2/3$ $\qquad$ $q_1 = 2/3$

We find that the optimal strategies of player A and player B in the original game are (0, 0, 2/3, 1/3) and (0, 0, 2/3, 1/3), respectively. The value of the game can be obtained by putting value of p_1 or q_1 in either of the expected payoff equations above. That is,

Expected gain to A $\qquad$ Expected loss to B

$4p_1 + 0.p_2 = 4(2/3) = 8/3$ $\qquad$ $4q_1 + 0_q2 = 4(2/3) = 8/3$

Here it may be noted that the expected loss to one store is the same as the expected gain to another store.

Example 7:

Two breakfast food manufacturers, ABC and XYZ are competing for an increased market share. The payoff matrix, shown in the following table, describes the increase in market share for ABC and decrease in market share of XYZ.

ABC	*XYZ*			
	Give Coupons	***Decrease Price***	***Maintain Present Strategy***	***Increase Advertising***
Give Coupons	*2*	*–2*	*4*	*1*
Decrease Price	*6*	*1*	*12*	*3*
Maintain Present Strategy	*–3*	*2*	*0*	*6*
Increase Advertising	*2*	*–3*	*7*	*1*

Determine optimal strategies for both the manufacturers and the value of the game.

Solution:

The first step is to search for a saddle point. There is no saddle point in the problem. The second step is to observe if the payoff matrix can be reduced in size by rules of dominance. Yes, each element of first row is less than the corresponding elements of second row. Therefore, first row is dominated by second row because payoffs are less attractive for ABC. Thus, after deleting first row, the reduced matrix becomes as shown below:

ABC	*XYZ*			
	B_1	B_2	B_3	B_4
A_2	6	1	12	3
A_3	–3	2	0	6
A_4	2	–3	7	1

In the reduced matrix, each element of fourth column is more than the corresponding element in second column. Therefore, fourth column is dominated by second column because payoffs are less attractive (more loss) for XYZ. Thus after deleting fourth column the reduced matrix becomes

ABC	XYZ		
	B_1	B_2	B_3
A_2	6	1	12
A_3	–3	2	0
A_4	2	–3	7

Further compare rows 1 and 3 and then columns 1 and 3 and delete the less attractive row and column from ABC's and XYZ's point of view. The reduced payoff matrix is shown below:

		XYZ	
ABC	*Give Coupons*	*Decrease Price*	
	B_1	B_2	*Probability*
Decrease Price, A_2	6	1	p_1
Maintain Present Strategy, A_3	–3	2	p_2
Probability	q_1	q_2	

The reduced 2 × 2 payoff matrix also does not have the saddle point. Thus, both ABC and XYZ use mixed strategies.

For ABC Let p_1 and p_2 be probabilities of selecting strategy A_3 (decrease price) and A; (maintain present strategy), respectively. Then the expected gain to ABC when XYZ uses its B_1 and B_2 strategies is given by

$$6p_1 - 3p_2 \text{ and } p_1 + 2p_2 \; ; \; p_1 + p_2 = 1$$

The probability p_1 and p_2 should be such that expected gains under both conditions are equal. That is,

$$6p_1 - 3p_2 = 3q_1 + 2q_2$$

$$6p_1 - 3(1 - p_1) = p_1 + 2(1 - p_1); \; p_1 + p_2 = 1$$

$$10p_1 = 5 \text{ or } p_1 = 1/2 \text{ and } p_2 = 1 - p_1 = 1/2$$

For XYZ Let q_1 and q_2 be probabilities of selecting strategies B_1 (give coupons) and B_2 (decrease price), respectively. Then the expected loss to XYZ when ABC uses its A_2 and A_3 strategies should be

$$6q_1 + q_2 = 3q_1 + 2q_2 \; ; \; q_1 + q_2 = 1$$

$$6q_1 + (1 - q_1) = -3q_1 + 2(1 - q_1)$$

$$10q_1 = 1 \text{ or } q_1 = 1/10 \text{ and } q_2 = 1 - q_1 = 9/10$$

Hence, optimal strategies for both the manufacturers are that ABC should adopt strategy A_2 (decrease price) 50 per cent of time and strategy A_3 (maintain present strategy) 50 per cent of time, while XYZ should adopt strategy 5, (give coupons) 10 per cent of time and strategy B_2 (decrease price), 90 per cent of time.

The expected gain and loss to ABC and XYZ can be calculated as shown below:

Expected gain to ABC : $6p_1 - 3p_2 = 6(1/2) - 3(1/2) = 3/2$

$p_1 + 2p_2 = 1/2 + 2(1/2) = 3/2$

Expected loss to XYZ : $6q_1 + q_2 = 6(1/10) + (9/10) = 3/2$

$-3q_1 + 2q_2 = -3(1/10) + 2(9/10) = 3/2$

Example 8:

XYZ Company manufactures parts for passenger cars and sells them in lots of 10,000 parts each. The company has a policy of inspecting each lot before it is actually shipped to the retailer. Five inspection categories, established for quality control, represent the percentage of defective items contained in each lot. These are given in the following table. The daily inspection chart for past 100 inspections shows the following rating or breakdown inspection:

The management is considering two possible courses of action:

(i) S_1: Shut down the entire plant operations and thoroughly inspect each machine.

Rating	***Proportion of Defective Items***	***Frequency***
Excellent (A)	*0.02*	*25*
Good (B)	*0.05*	*30*
Acceptable (C)	*0.10*	*20*
Fair (D)	*0.15*	*20*
Poor (E)	*0.20*	*5*
		Total = 100

(ii) S_2: Continue production as it now exists but offer the customer a refund for defective items that are discovered and subsequently returned.

The first alternative will cost Rs. 600 while the second alternative will cost the company Re 1 for each defective item that is returned. What is the optimum decision for the company? Find the EVPI.

Solution:

Calculations of inspection and refund cost are shown in Table 5.3.

Table 5.3 : Inspection and Refund Cost

Rating	***Defective***	***Probability Rate***	***Cost***		***Opportunity Loss***	
			Inspect	***Refund***	***Inspect***	***Refund***
A	0.02	0.25	600	200	400	0
B	0.05	0.30	600	500	100	0
C	0.10	0.20	600	1,000	0	400
D	0.15	0.20	600	1,500	0	900
E	0.20	0.05	600	2,000	0	1,400
		1.00	600*	670	170*	240

The cost of refund is calculated as follows:

For lot A : $10{,}000 \times 0.02 \times 1.00 = \text{Rs. } 200$

Similarly, the cost of refund for other lots is calculated.

Expected cost of refund is:

$200 \times 0.25 + 500 \times 0.30 + ... + 2{,}000 \times 0.05 = \text{Rs. } 670$

Now expected cost of inspection is:

$600 \times 0.25 + 600 \times 0.30 + ... + 600 \times 0.05 = \text{Rs. } 600$

Since the cost of refund is more than the cost of inspection, the plant should be shut down for inspection. Also,

$$\text{EVPI} = \text{EOL of inspection} = \text{Rs. } 170.$$

Example 9:

The demand pattern of the cakes made in a bakery is as follows:

No. of cakes demanded	:	*0*	*1*	*2*	*3*	*4*	*5*
Probability	:	*0.05*	*0.10*	*0.25*	*0.30*	*0.20*	*0.10*

If the preparation cost is Rs. 3 per unit and selling price is Rs. 4 per unit, how many cakes should the baker bake to maximize profit?

Solution:

Given that incremental cost (IC) is Rs. 3 per unit and incremental price (IP) Rs. 4 per unit, the minimum required probability of selling at least an additional unit of cake to justify the stocking of that unit is given by

$$p = \frac{IC}{IC + IP} = \frac{3}{3+4} = 0.75$$

This probability means that the bakery owner must have demand level k such that p (demand $\geq$ k) $\geq$ p.

The cumulative probabilities of greater than type are computed as shown in Table 5.4.

Table 5.4

Demand (No. of cakes)	***Probability P (Demand = k)***	***Cumulative Probability P (Demand ≥ k)***
0	0.05	1.00
1	0.10	0.95
2	0.25	0.85
3	0.30	0.60
4	0.20	0.30
5	0.10	0.10

Since the highest value of A for which p(demand ≥ k) exceeds the critical ratio p = 0.75 is k = 2, the optimal decision is to prepare only two cakes.

Example 10:

A manager must choose between two investments A and B which are calculated to yield net profits of Rs. 1,200 and Rs. 1,600, respectively with probabilities subjectively estimated at 0.75 and 0.60. Assume the manager's utility function reveals that utilities for Rs. 1,200 and Rs. 1,600 are 45 and 50 utils, respectively. What is the best choice on the basis of expected utility value (EUV)?

Solution:

The expected utility value is expressed as:

$$EUV = \sum_{i=1}^{m} U_i p_i$$

where U_1 = utility value of state of nature i

p_i = probability value of state of nature i

Then EUV (A) = $p_A\ U_A = 0.75 \times 45 = 33.75$ utils

EUV (B) = $p_B U_B = 0.60 \times 50 = 30$ utils

Since EW(A) > EUV(B), therefore, best choice is the investment A.

Example 11:

Two Firms A and B have for years been selling a competing product which forms a part of both firms' total sales. The marketing executive of Firm A raised the question: 'What should be the firm's strategies in terms of advertising for the product in question'. The market research team of Firm A developed the following data for varying degree of advertising:

(i) No advertising, medium advertising, and large advertising for both firms will result in equal market shares.

(ii) Firm A with no advertising: 40 per cent of the market with medium advertising by Firm B and 28 per cent of the market with large advertising by Firm B.

(iii) Firm A using medium advertising: 70 per cent of the market with no advertising by Firm B and 45 per cent of the market with large advertising by Firm B.

(iv) Firm A using large advertising: 75 per cent of the market with no advertising by Firm B and 47.5 per cent of the market with medium advertising by Firm B.

(a) *Based upon the foregoing information, answer the marketing executive's questions:*

(b) *What advertising policy should Firm A pursue when consideration is given to the above factors: selling price, Rs. 4.00 per unit; variable cost of product, Rs. 2.50 per unit; annual volume of 30,000 units for Firm A, cost of annual medium advertising Rs. 5,000 and cost of annual large advertising Rs. 15,000? What contribution, before other fixed costs, is available to the firm?*

Solution:

The payoff matrix of the game between Firms A and B is as follows:

		Firm B		
Firm A	***No. Advt., B_1***	***Medium Advt., B_2***	***Large Advt., B_3***	***Row minimum***
No Advt., A_1	50	40	28	28
Medium Advt., A_2	70	50	45	45
Large Advt., A_3	75	47.5	50	47.5
Column maximum	75	50	50	

From the payoff matrix it is observed that there is no saddle point in the problem.

Applying rules of dominance, delete first row (dominated by third row) and then first column (dominated by both columns 2 and 3) from the payoff matrix. The reduced payoff matrix so obtained is shown below:

	Firm B		
Firm A	B_2	B_3	*Probability*
A,	50	45	P_1
A,	47.5	50	P_2

The reduced 2 × 2 payoff matrix also does not have the saddle point. Thus, both the firms use mixed strategies. Adopting the same procedure as discussed in earlier examples, the expected gain to Firm A can be calculated as follows:

Expected Gain to Firm A

$$50p_1 + 47.5p_2 = 45p_1 + 50p_2 \; ; \; p_1 + p_2 = 1$$

$$50p_1 + 47.5(1 - p_1) = 45p_1 + 50(1 - p_1)$$

$$7.5p_1 = 2.5 \text{ or } p_1 = 1/3 \text{ and } p_2 = 1 - p_1 = 2/3.$$

Expected gain $= 50p_1 + 47.5p_2$

$= 50(1/3) + 47.5(2/3) = 145/3.$

Thus, the optimal policy for Firm A is to apply strategy A_2 (medium advertising) with probability 0.33 and strategy Ay (large advertising) with probability 0.67 on any one play of the game. With this policy, the firm may expect to gain 145/3 = 48.3 per cent of the market share.

Table 5.5 Market Share of Firm A

		Firm B	
Firm A	*No Advt.*	*Medium Advt.*	*Large Advt.*
No Advt.	0.50×30,000=15,000	0.40×30,000=12,000	0.28×30,000=8,400
Medium Advt.	0.70×30,000=21,000	0.50×30,000=15,000	0.45×30,000=13,500
Large Advt.	0.75×30,000=22,500	0.475×30,000=14,250	0.50×30,000=15,000

Given that the expenditure on medium and large advertisements is Rs. 5,000 and Rs. 15,000, respectively, net profit to Firm A can be calculated by using following equation:

Net profit = (Sales price – Cost price) × Sales volume – Advertising expenditure The net profit to Firm A is shown in Table 5.6.

Table 5.6 Profit to Firm A

		Finn B	
Firm A	*No Advt.*	*Medium Advt.*	*Large Advt.*
No advt.	1.5×15,000 = 22,500	1.5×12,000 =18.000	1.5×8,400 = 12.600
Medium advt.	1.5×21,000–5,000 = 26,500	1.5×15.000–5.000 = 17.500	1.5×13,500–5,000 = 15,250
Large advt.	1.5×22,500 – 15,000 = 18,750	1.5×14.250 – 15,000 = 6.375	1.5×15,000- 15,000 = 7,500

Observations

1. If Firm A chooses the strategy of *'No advertising'*, then minimum profit is Rs. 12,600, because Firm B can adopt its strategy 'Large advertising'.
2. If Firm A chooses the strategy of *'Medium advertising'*, then minimum profit is Rs. 15,250 because Firm B can again adopt its strategy *'Large advertising'*.
3. If Firm A chooses the strategy of *'Large advertising'*, then minimum profit is Rs. 6,375 because firm B can adopt its strategy *'Medium advertising'*.

Based on these observations, the Firm A must adopt the policy of 'medium advertising' to gain maximum profit of Rs. 15,250 among these three alternatives and must spend Rs. 5,000 for advertising.

Arithmetic Method

The arithmetic method (also known as short cut method) provides an easy method for finding optimal strategies for each player in a payoff matrix of size 2 × 2 without saddle point. The steps of this method are as follows:

1. Find the differences between the two values in the first row and put it against second row of the matrix, neglecting negative sign (if any).
2. Find the difference between the two values in the second row and put it against first row of the matrix, neglecting negative sign (if any).
3. Repeat Steps 1 and 2 for the two columns also.

The values so obtained by 'swapping the differences' represent the optimal relative frequencies of play for both players strategies. These may be converted to probabilities by dividing each of them by their sum. The value of the game can be obtained by applying any of the methods discussed earlier.

Graphical Method

The graphical method is useful for the game where the payoff matrix is of the size 2 × n or m × 2, *i.e.*, the game with mixed strategies that has only two undominated pure strategies for one of the players in the two-person zero-sum game.

Optimal strategies for both the players assign no-zero probabilities to the same number of pure strategies. Therefore, if one player has only two strategies, the other will also use the same number of strategies. Hence, this method is useful in finding out which of the two strategies can be used.

Consider the following 2 × n payoff matrix of a game without saddle point.

		Player B			
Player A	B_1	B_2	...	B_n	*Probability*
A_1	a_{11}	a_{12}	...	a_{1n}	p_1
A_2	a_{21}	a_{22}	...	a_{2n}	p_2
Probability	q_1	q_2	...	q_n	

Player A has two strategies A_1 and A_2 with probability of their selection p_1 and p_2, respectively, such that $p_1 + p_2 = 1$ and $p_1, p_2 \geq 0$. Now for each of the pure strategies available to player B, expected pay off for player A would be as follows:

B's Pure Strategies	*A's Expected Pay off*
B_1	$a_{11}\ p_1 + a_{21}\ p_2$
B_2	$a_{12}\ p_1 + a_{22}\ p_2$
$\vdots$	$\vdots$
B_n	$a_{1n}\ p_1 + a_{2n}\ p_2$

According to the maximin criterion for mixed strategy games, player A should select the value of probability p_1 and p_2 so as to maximize his minimum expected payoffs. This may be done by plotting the straight lines representing player A's expected payoff values.

The highest point on the lower boundary of these lines will give maximum expected payoff among the minimum expected payoffs and the optimum value of probability p_1 and p_2.

Now the two strategies of player B corresponding to those lines which pass through the maximin point can be determined. It helps in reducing the size of the game to (2×2), which can be easily solved by any of the methods discussed earlier.

The $(m \times 2)$ games are also treated in the same way except that the upper boundary of the straight lines corresponding to B's expected payoff will give the maximum expected payoff to player B and the lowest point on this boundary will then give the minimum expected payoff (minimax value) and the optimum value of probability q_1 and q_2.

Example 12:

Use graphical method in solving the following game and find the value of the game.

		Player B		
Player A	***B_1***	***B_2***	***B_3***	***B_4***
A_1	*22*	*3*	*–2*	
A_2	*4*	*3*	*2*	*6.*

Solution:

The game does not have a saddle point. If the probability of player A's playing A_1 and A_2 in the strategy mixture is denoted by p_1 and p_2, respectively, where $p_2 = 1 - p_1$, then the expected payoff (gain) to player A will be

B's Pure Strategies	*A's Expected Payoff*
B_1	$2\ p_1 + 4\ p_2$
B_2	$2\ p_1 + 3\ p_2$
B_3	$3\ p_1 + 2\ p_2$
B_4	$-2p_1 + 6p_2$

These four expected payoff lines can be plotted on the graph to solve the game.

The graph for player A : A graphic solution is shown in Fig. 12.2. Here, the probability of player A's playing A_1, *i.e.,* p, is measured on the x-axis. Since p_1 cannot exceed 1, the x-axis is cut-off at $p_1 = 1$. The expected payoff of player A is measured along y-axis. From the game matrix, if player B plays B_1 the expected payoff of player A is 2 when A plays A_1 with $p_1 = 1$ and 4 when A plays A_2 with $p_1 = 0$. These two extreme points are connected by a straight line, which shows the expected payoff of A when B plays B_1 Three other straight lines are similarly drawn for B_2, B_3 and B_4.

It is assumed that player B will always play his best possible strategies yielding the worst result to player A. Thus, the payoffs (gains) to A are represented by the lower boundary when he is faced with the most unfavourable situation in the game. Since player A must choose his best possible strategies in order to realize a maximum expected gain, the highest expected gain is found at point P, where two straight lines

$$E_3 = 3p_1 + p2_2 = 3p_1 + 2(1 - p_1)$$

$$E_4 = -2p_1 + 6p_2 = -2p_1 + 6(1 - p_1)$$

meet. In this manner the solution to the original (2 × 4) game reduces to that of the game with payoff matrix of size (2 × 2) as given below:

	Player B	
Player A	B_3	B_4
A_1	2	6

The optimum payoff to player A can now be obtained by setting E_3 and E_4 equal and solving for p_1 *i.e.,*

$$3p_1 + 2(1 - p_1) = -2p_1 + 6(1 - p_1)$$

or $$p_1 = 4/9;\ p_2 = 1 - p_1 = 5/9$$

Substituting the value of p_1 and p_2 in the equation for E_3 (or E_4) we have,

Value of the game, $V = 3 \times 4/9 + 2 \times 5/9 = 22/9$

The optimal strategy mix of player B can also be found in the same manner as for player A. If the probabilities of B's selecting strategy B_3 and B_4 are denoted by q_3 and q_4, respectively, then the expected loss to B will be

$$L_3 = 3q_3 - 2q_4 = 3q_3 - 2(1 - q_3) \text{ (if A selects } A_1\text{)}$$

$$L_4 = 2q_3 + 6q_4 = 2q_3 + 6(1 - q_3) \text{ (if A selects } A_2\text{)}$$

To solve for q_3, equate the two equations

$$3q_3 - 2(1 - q_3) = 2(73 + 6(1 - q_3)$$

or $$q_3 = 8/9;\ q_4 = 1 - q_3 = 1/9$$

Substituting the value of q_3 and q_4 in the equation for L_3 (or L_4), we have

Value of the game, $V = 3 \times 8/9 - 2 \times 1/9 = 22/9$

Example 13:

Obtain the optimal strategies for both persons and the value of the game for two-person zero-sum game whose payoff matrix is as follows:

	Player B	
Player A	B_1	B_1
A_1	1	–3
A_2	3	5
A_3	–1	6
A_4	4	1
A_5	2	2
A_6	–5	0.

Solution:

The game does not have any saddle point. If the probability of player B's playing strategies B_1 and B2 in the strategy mix is denoted by q_1 and q_2 such that $q_1 + q_2 = 1$, then the expected payoff to player B will be:

A's Pure Strategies	***B's Expected Payoff***
A_1	$q_1 - 3q_2$
A_2	$3q_1 + 5q_2$
A_3	$-q_1 + 6q_2$
A_4	$4q_1 + q_2$
A_5	$2q_1 + 2q_2$
A_6	$-5q_1 + 0q_2$

The six expected payoff lines can be plotted on the graph to solve the game.

The graph for player B : A graphic solution is shown in Fig. 12.3 where the probability of player B is playing B_1 *i.e.*, q_1 is measured on the x-axis. Since q_1 cannot exceed 1, therefore x-axis is cut off at $q_1 = 1$. The expected payoff of player B is measured along y-axis. From the game matrix, if player A plays A_1, the expected payoff of player B is 1 when he plays B_1 with $q_1 = 1$ and –3 when he plays B_2 with $q_1 = 0$. These two extreme points are connected by a straight line, which shows the expected payoff to B when A plays A_1. Five other straight lines are similarly drawn for A_2 to A_6.

It is assumed that player A will always play his best possible strategies yielding the worst result to player B. Thus, payoffs (losses) to B are represented by the upper boundary when he is faced with the most unfavourable situation in the game. According to the minimax criterion, player B will always select a combination of strategies B_1 and B_2, so that he minimizes the losses. In this case also the optimum, solution occurs at the intersection of the two payoff lines.

$$E_3 = 3q_1 + 5q_2 = 3q_1 + 5(1 - q_1)$$
$$E_4 = 4q_1 + q_2 + (1 - q_1)$$

The solution to the original (6 × 2) game reduces to that of the game with payoff matrix of size (2 × 2) as shown below:

	Player B	
Player A	B_1	B_2
A_2	3	5
A_4	4	1

Now using the usual method of solution for a (2 × 2) game, the optimum strategies can be obtained as given below:

Player A : (0, 3/5, 0, 2/5, 0, 0) ; Player B : (4/5, 1/5)

and, Value of the game, V = 17/5.

Example 14:

The probability of the demand for lorries for hiring on any day in a given district is as follows:

No. of lorries demanded	:	*0*	*1*	*2*	*3*	*4*
Probability	:	*0.1*	*0.2*	*0.3*	*0.2*	*0.2*

Lorries have a fixed cost of Rs. 90 each day to keep the daily hire charges (net of variable costs of running) Rs. 200. If the lorry-hire company owns 4 lorries, what is its daily expectation? If the company is about to go into business and currently has no lorries, how many lorries should it buy?

Solution:

It is given that Rs. 90 is the fixed cost and Rs. 200 is variable cost. Now the payoff values with 4 lorries at the dispcsal of decision-maker are calculated as under:

No. of Lorries demanded	:	0	1	2	3	4
Payoff	:	0–90×4	200–90×4	400–90×4	600–90×4	800–90×4
(with 4 lorries)		= –360	= –160	= 40	= 240	= 440

Thus daily expectation is obtained by multiplying the payoff values with the given corresponding probabilities of demand:

Daily Expectation = (–360)(0.1) + (–160)(0.2) + (40)(0.3) + (240)(0.2) + (440)(0.2) = Rs. 80. The conditional payoffs and expected payoffs for each course of action are shown in Tables 5.7 and 5.8.

Table 5.7 : Conditional Payoff Values

Demand of Lorries	*Probability*	*Conditional Payoff (Rs.) due to Decision to Purchase Lorries (Course of Action)*				
		0	1	2	3	4
0	0.1	0	–90	–180	–270	–360
1	0.2	0	110	20	–70	–160
2	0.3	0	110	220	130	40
3	0.2	0	110	220	330	240
4	0.2	0	110	220	330	440

Table 5.8 : Expected Payoffs and EMV

Demand of Lorries	*Probability*	*Conditional Payoff (Rs.) due to Decision to Purchase Lorries (Course of Action)*				
		0	1	2	3	4
0	0.1	0	–9	–18	–27	–36
1	0.2	0	22	4	–14	–32
2	0.3	0	33	66	39	12
3	0.2	0	22	44	66	48
4	0.2	0	22	44	66	88
EMV		0	90	140	130	80

Since EMV Rs. 140 for the course of action 2 is highest, the company should buy 2 lorries.

Example 15:

An equipment costs Rs. 1,20,000 when installed. Old equipment can be sold for Rs. 20.000. The equipment will yield about Rs. 50,000 per year. If the corporate income lax rate is 50%. determine the payback period.

Solution:

$$\text{Payback period} = \frac{\text{Net investment}}{\text{Annual after - tax earnings}}$$

$$= \frac{\text{Rs. } (1{,}20{,}000 - 20{,}000)}{\text{Rs. } (50{,}000 \times 0.5)} = 4 \text{ years.}$$

Example 16:

A project requires a cash outlay of Rs. 60,000 and yields an annual cash inflow of Rs. 12.000 for 7 years. What is its payback period?

Solution:

$$\text{Payback period} = \frac{\text{Rs. } 60{,}000}{\text{Rs. } 12{,}000} = 5 \text{ years.}$$

Example 17:

A company manufactures goods for a market in which the technology of the product is changing rapidly. The research and development department has produced a new product which appears to have potential for commercial exploitation. A further Rs. 60,000 is required for development testing.

The company has 100 customers and each customer might purchase at the most one unit of the product. Market research suggests that a selling price of Rs. 6,000 for each unit with total variable costs of manufacturing and selling estimate are Rs. 2,000 for each unit.

From previous experience, it has been possible to derive a probability distribution relating to the proportion of customers who will buy the product as follows:

Proportion of customers	:	*0.04*	*0.08*	*0.12*	*0.16*	*0.20*
Probability	:	*0.10*	*0.10*	*020*	*0.40*	*020*

Determine the expected opportunity losses, given no other information than that stated above, and state whether or not the company should develop the product.

Solution:

If p is the proportion of customers who purchase the new product, the conditional profit is:

$(6{,}000 - 2{,}000) \times 100\,p - 60{,}000 = \text{Rs. } (4{,}00{,}000\,p - 60{,}000)$

Let N_i ($i = 1, 2, ..., 5$) be the possible states of nature, *i.e.*, proportion of the customers who will buy the new product and S_1 (develop the product) and S_2 (do not develop the product) be the two courses of action.

The conditional profit values (payoffs) for each pair of A's and S's are shown in Table 5.9.

Table 5.9 : Conditional Profit Values (Payoffs)

State of Nature (Proportion of Customers, p)	*Conditional Profit = Rs. (4,00,000 p – 60,000) Course of Action*	
	S_1 ***(Develop)***	S_2 ***(Do not Develop)***
0.04	–44,000	0
0.08	–28,000	0
0.12	–12,000	0
0.16	4,000	0
0.20	20,000	0

Opportunity loss values are shown in Table 5.10.

Table 5.10 : Opportunity Loss Values

State of Nature	*Probability*	*Conditional Profit (Rs.)*		*Opportunity Loss (Rs.)*	
		S_1	S_2	S_1	S_2
0.04	0.1	–44,000	0	44,000	0
0.08	0.1	–28,000	0	28,000	0
0.12	0.2	–12,000	0	12,000	0
0.16	0.4	4,000	0	0	4,000
0.20	0.2	20,000	0	0	20,000

Using the given estimates of probabilities associated with each state of nature, the expected opportunity loss (EOL) for each course of action is given below:

EOL (S_1) = 0.1 (44,000) + 0.1 (28,000) + 0.2 (12,000) + 0.4 (0) + 0.2 (0)

= Rs. 9,600

EOL (S_2) = 0.1 (0) + 0.1 (0) + 0.2 (0) + 0.4 (4,000) + 0.2 (20,000)

= Rs. 5,600.

Since the company seeks to minimize the expected opportunity loss, the company should select course of action S_2 (do not develop the product) with minimum EOL.

Example 18:

A company needs to increase its production beyond its existing capacity. It has narrowed the alternatives to two approaches to increase the production capacity: (a) expansion, at a cost of Rs. 8 million, or (b) modernization at a cost of Rs. 5 million. Both approaches would require the same amount of time for implementation. Management believes that over the required payback period, demand will either be high or moderate. Since high demand is considered to be somewhat less likely than moderate demand, the probability of high demand has been set at 0.35. If the demand is high, expansion would gross an estimated additional Rs. 12 million but modernization only an additional Rs. 6 million, due to lower maximum production capability. On the other hand, if the demand is moderate, the comparable figures would be Rs. 7 million for expansion and Rs. 5 million for modernization.

(a) Calculate the conditional profit in relation to various action-and-outcome combinations and states of nature.

(b) If the company wishes to maximize its expected monetary value (EMV), should it modernize or expand?

(c) Calculate the EVPI.

(d) Construct the conditional opportunity loss table and also calculate EOL.

Solution (a):

Defining the states of nature: High demand and Moderate demand (over which the company has no control) and courses of action (company's possible decisions): Expand and Modernize.

Since the probability that the demand is high estimated at 0.35, the probability of moderate demand must be (1 – 0.35) = 0.65. The calculations for conditional profit values are shown in Table 5.11.

Table 5.11 : Conditional Profit Table

State of Nature (Demand)	*Conditional Profit (million Rs.) due to Course of Action*	
	Expand (S_1)	*Modernize (S_2)*
High demand (N_1)	12 – 8 = 4	6 – 5 = 1
Moderate demand (N_2)	7 – 8 = –1	5 – 5 = 0

(b) The payoff table (Table 5.11) can be rewritten as follows along with the given probabilities of states of nature.

Table 5.12 : Payoff Table

State of Nature (Demand)	*Probability*	*Conditional Profit (million Rs.) due to Course of Action*	
		Expand	*Modernize*
High demand	0.35	4	1
Moderate demand	0.65	–1	0

The calculation of EMV for each course of action S_1 and S_2 is given below:

$$EMV(S_1) = 0.35(4) + 0.65(-1) = \text{Rs. } 0.75 \text{ million}$$

$$VMV(S_2) = 0.35(1) + 0.65(0) = \text{Rs. } 0.35 \text{ million}$$

Thus to maximize EMV, the company must choose course of action S_1 (expand). The EMV of the optimal course of action is generally denoted by EMV*. Therefore,

$$EMV^* = EMV(S)) = \text{Rs. } 0.75 \text{ million}$$

(c) To calculate EVPI, we shall first calculate EPPI. For calculating EPPI, we choose optimal course of action for each state of nature, multiply its conditional profit by the given probability to get weighted profit, and then sum these weights as shown in Table 5.13.

Table 5.13

State of Nature (Demand)	*Probability*	*Optimal Course of Action*	*Profit from Optimal Course of Action (Rs. million)*	
			Conditional Profit	*Weighted Profit*
High demand	0.35	S_1	4	4 × 0.35 = 1.40
Moderate demand	0.65	S_2	0	0 × 0.65 = 0
				EPPI = 1.40

The optimal EMV* is Rs. 0.75 million corresponding to the course of action S_1. Then

$$\text{EVPI} = \text{EPPI} - \text{EMV}(S_1)$$
$$= 1.40 - 0.75 = \text{Rs. } 0.65 \text{ million.}$$

In other words, if the company could get a perfect information (or forecast) of demand (high or moderate), it should consider paying up to Rs. 0.65 million for an information.

The expected value of perfect information in business helps in getting an absolute upper bound on the amount that should be spent to get additional information on which to base a given decision.

(d) The opportunity loss values are shown in Table 5.14.

Table 5.14 : Conditional Opportunity Loss Table

State of Nature (Demand)	*Probability*	*Conditional Profit (Rs. million) due to Course of Action*		*Conditional Opportunity Loss (Rs. million) due to Course of Action*	
		S_1	S_2	S_1	S_2
High demand (N_1)	0.35	4	1	0	0
Moderate demand (N_2)	0.65	–1	0	1	0

The conditional opportunity loss values may be explained as follows: If high demand (N_1) occurred, then the maximum profit of Rs. 4 million would be achieved by selecting course of action S_1. Therefore, the selection of S_1 would result in zero opportunity loss, as it is the best decision that can be made if N_1 occurs. If course of action S_2 was chosen with a payoff of one million, then this would result in an opportunity loss of 4 – 1 = Rs. 3 million. If moderate demand (N_2) occurred, then the best course of action would be S_2 with Rs. zero million profit. Thus, opportunity loss would be associated with the selection of S_2 but if S_1 was selected, then the opportunity loss would be 0 – (–1) = Rs. 1 million. That is, the company would have been Rs. 1 million worse off if it had chosen course of action S_2.

Using the given estimates of probabilities associated with each state of nature, *i.e.*, $P(N_1) = 0.35$, and $P(N_2) = 0.65$, the expected opportunity losses for the two courses of action are:

$$\text{EOL}(S_1) = 0.35(0) + 0.65(1) = \text{Rs. } 0.65 \text{ million}$$
$$\text{EOL}(S_2) = 0.35(3) + 0.65(0) = \text{Rs. } 1.05 \text{ million.}$$

Since decision-maker seeks to minimize the expected opportunity loss, he must select course of action S_1 as it produces the smallest expected opportunity loss.

Example 19:

Calculate the payback period fur a project which requires a cash outlay Rs. 30.000 and generates cash inflows of Rs. 11,000, Rs. 10,000. Rs. 7,000 and Rs. 6,000.

Solution:

When we add up the cash inflows, we find that Rs. 28,000 of the original outlay is recovered in the first three years. In the fourth year the balance Rs. 2,000 of the original outlay needs to be recovered and the cash inflow during the year is Rs. 6,000. Assuming that the cash inflows occur uniformly during the year. the time required to recover Rs. 2,000 will be

$$\frac{\text{Rs. } 2{,}000}{\text{Rs. } 6{,}000} \times 12 = 4 \text{ months.}$$

The payback period is 3 years and 4 months.

Example 20:

Net investment for a new equipment is Rs. 1,00,000. Based on ten-year expectation, average earnings per year, after taxes and depreciation are estimated at Rs. 20,000. Find average rate of return.

Solution:

$$\text{Average rate of return} = \frac{\text{Rs.} 20{,}000}{\text{Rs. } 1{,}00{,}000} \times 100 = 20\%.$$

Example 21:

The following matrix gives the payoff of different strategies (alternatives) S_1, S_2, S_3 against conditions (events) N_1, N_2, N_3 and N_4:

Table 5.15

	N_1	N_2	N_3	N_4
S_1	*Rs. 4,000*	*Rs. –100*	*Rs. 6,000*	*Rs. 18,000*
S_2	*20,000*	*5,000*	*400*	*0*
S_3	*20,000*	*15,000*	*–2,000*	*1,000*

Indicate the decision taken under the following approach :

(i) Pessimistic

(ii) Optimistic

(iii) Regret and

(iv) Equal probability.

Solution:

For the given payoff matrix, the values corresponding to pessimistic, optimistic and equal penalty criteria are given as below.

Table 5.16

	Pessimistic (maximin) value	*Optimistic (maximam) value*	*Equal probability value* $= 1/n\ (P_1 + P_2 + ... + P_4)$
S_1	–Rs. 100	Rs. 18,000	Rs. 1/4 (4,000–100+6,000+18,000) = Rs. 6,975
S_2	Rs. 0	Rs. 20,000	Rs. 1/4 (20,000+5,000+400+0) = Rs. 6,350
S_3	–Rs. 2,000	Rs. 20,000	Rs. 1/4 (20,000+15,000–2,000+1,000) = Rs. 8,500

Thus under pessimistic approach, S_2 is the optimal decision, under optimistic approach, S_2 or S_3 are the decision alternatives and under equal probability approach, S_3 is the alternative to be selected.

Table 5.17 represents the regret for every event and for each alternative calculated by the expression.

ith regret = (maximum payoff – ith payoff) for the jth event.

Table 5.17

	N_1 *Regret (Rs.)*	N_2 *Regret (Rs.)*	N_3 *Regret (Rs.)*	N_4 *Regret (Rs.)*	*Maximum regret (Rs.)*
S_1	16,000	15,100	0	0	16,000
S_2	0	10,000	5,600	18,000	18,000
S_3	0	0	8,000	17,000	17,000

The decision alternative S_1 would be chosen since it corresponds to the minimal of the maximum possible regrets.

Example 22:

Given the following payoff function for each act a_1 and a_2.

$$Q_1 = -25 + 40x,$$

$$Q_2 = -80 + 29x,$$

(i) *Find the break even value of x.*

(ii) *If x = 5, which is the better act?*

(iii) *If x = 5, what is the regret of poor strategy?*

(iv) If x = –10, which is the better act?

(v) If x = –10, what is the regret of poor strategy?

Solution:

(i) For break even point we have,

Q_{a_1} Q_{a_2}

$\Rightarrow -25 + 40x = -80 + 29x$

$\Rightarrow 11x = -55$ or $x = -5$

$\therefore$ –5 is the break even point.

(ii) when $x = 5$, $Q_{a_1} = -25 + 200 = 175$,

$Q_{a_2} = -80 + 145 = 65$.

$\therefore$ a_1 is the better act.

(iii) For $x = 5$, the regret of the poor strategy $a_2 = 175 - 65 = 110$.

(iv) When $x = -10$, $Q_{a_1} = -25 - 400 = -425$,

$Q_2 = -80 - 290 = -370$.

$\therefore$ a_2 is the better act.

(v) For $x = -10$, the regret of the poor strategy $a_1 = -370 + 425 = 55$.

Example 23:

A newspaper boy has the following probabilities of selling a magazine:

No. of copies sold	***Probability***
10	*0.10*
11	*0.15*
12	*0.20*
13	*0.25*
14	*0.30*
	1.00

Cost of copy is 30 paise and sale price is 50 paise. He cannot return unsold copies. How many copies should he order?

Solution:

The no. of copies for purchases and for sales which have meaning to the newsboy are 10, 11, 12, 13, or 14. These are his sales magnitudes. There is no reason for him to buy less than 10 or more than 14 copies. Table 5.18, *the conditional profit table,* shows the profit resulting from any possible combination of supply and demand. Stocking of 10 copies each day will always result in a profit of 200 paise irrespective of the demand. For instance, even if the demand on some day is 13 copies, he can sell only 10 and hence his conditional profit is 200 paise.

When he stocks 11 copies, his profit will be 220 paise on days when buyers request 11, 12, 13, or 14 copies. But on days when he has 11 copies on stock and buyers buy only 10 copies, his profit decreases to 170 paise. The profit of 200 paise on the 10 copies sold must be reduced by 30 paise the cost of one copy left unsold. The same will be true when he stocks 12, 13 or 14 copies. Thus, the conditional profit in paise is given by

$$\text{Payoff} = 20 \times \text{copies sold} - 30 \times \text{copies unsold.}$$

Table 5.18 : Conditional Profit Table (Paise)

Possible demand (no. of copies)	*Probability*	*Possible stock action*				
		10 copies	*11 copies*	*12 copies*	*13 copies*	*14 copies*
10	0.10	200	170	140	110	80
11	0.15	200	220	190	160	130
12	0.20	200	220	240	210	180
13	0.25	200	220	240	260	230
14	0.30	200	220	240	260	280

Next, the expected value of each decision alternative is obtained by multiplying its conditional profit by the associated probability and adding the resulting values. This is shown in Table 5.19.

Table 5.19 : Expected Profit Table

Possible damand	*Probability*	*Expected profit from stocking (paise)*				
		10 copies	*11 copies*	*12 copies*	*13 copies*	*14 copies*
10	0.10	20	17	14	11	8
11	0.15	30	33	28.5	24	19.5
12	0.20	40	44	48	42	36
13	0.25	50	55	60	65	57.5
14	0.30	60	66	72	78	84
Total expected profit (paise)		200	215	222.5	220	205

The newsboy must, therefore, order 12 copies to earn the highest possible average daily profit of 222.5 paise. This stocking will maximize the total profits over a period of time. Of course there is no guarantee that he will make a profit of 222.5 paise tomorrow. However , if he stocks 12 copies each day under the conditions given, he will have average profit of 222.5 paise per day. This is the best he can do because the choice of any one of the other four possible stock actions will result in a lower daily profit.

Example 24:

Solve example 23 by EOL criterion.

Solution:

Table 5.20 : Conditional Profit Table (Paise)

Possible demand (no. of copies) (Event)	*Probability*	*Possible stock action (alternative)*				
		10 copies	*11 copies*	*12 copies*	*13 copies*	*14 copies*
10	0.10	200	170	140	110	80
11	0.15	200	220	190	160	130
12	0.20	200	220	240	210	180
13	0.25	200	220	240	260	230
14	0.30	200	220	240	260	280

The best alternative for demand of 10 copies is to order 10 copies resulting in optimal profit of 200 paise. The conditional opportunity loss for each stock action (alternative) for this event is obtained just by subtracting the respective conditional profits from 200 paise. Likewise, for demand of 11, 12, 13 and 14 copies subtract the conditional payoff values for each of these rows from the maxima of that row. The resulting conditional opportunity loss (COL) table is shown below.

Table 5.21 : Conditional Loss Table (Paise)

Possible demand (no. of copies) (Event)	*Probability*	*Possible stock action (alternative)*				
		10 copies	*11 copies*	*12 copies*	*13 copies*	*14 copies*
10	0.10	0	30	60	90	120
11	0.15	20	0	30	60	90
12	0.20	40	20	0	30	60
13	0.25	60	40	20	0	30
14	0.30	80	60	40	20	0

Expected opportunity loss (EOL), can now be computed by multiplying the probability of each of state of nature with the appropriate loss value and adding the resulting products. For instance for holding a stock of 10 copies,

$$EOL = 0.10 \times 0 + 0.15 \times 20 + 0.20 \times 40 + 0.25 \times 60 + 0.30 \times 80$$
$$= 0 + 3 + 8 + 15 + 24 = 50 \text{ paise.}$$

EOL values for various stock actions are computed in table 9.14.

Table 5.22 : Expected Loss Table (Paise)

Possible demand (no. of copies) (Event)	*Probability*	*Possible stock action (alternative)*				
		10 copies	*11 copies*	*12 copies*	*13 copies*	*14 copies*
10	0.10	0	3	6	9	12
11	0.15	3	0	4.5	9	13.5
12	0.20	8	4	0	6	12
13	0.25	15	10	5	0	7.5
14	0.30	24	18	12	6	0
EOL (paise)		50	35	27.5	30	45

The optimum stock action is the one which will minimize expected opportunity losses; this action calls for the stocking of 12 copies each day, at which point there is minimum expected loss of 27.5 paise.

Example 25:

An ice cream retailer buys ice-cream at a cost of Rs. 5 per cup and sells it for Rs. 8 per cup; any remaining unsold at the end of the day can be disposed of at a salvage price of Rs. 2 per cup. Past sales have ranged between 15 and 18 cups per day ; there is no reason to b have that sales volume will take on any other magnitude in future. Find the EVM if the sale history has the following probabilities:

Market Size	:	*15*	*16*	*17*	*18*
Probability	:	*0.10*	*0.20*	*0.40*	*0.30*

Solution:

From the data given in the problem, we can calculate the conditional profit values for each stock action and event (market size) combination. If CP denotes the conditional profit, S the quantity in stock, and D the market demand, then

$$CP = (8 - 5)\,S = 3S, \text{ when } D > S,$$
$$8D - 5S + 2\,(S - D)\,, \text{ when } D < S.$$

The resulting pay off matrix is given below. A stock of 15 cups each day will result in daily profit of Rs. 45 irrespective of the demand. A stock of 16 cups each day will result in a profit of Rs. 48 when the demand is 16 cups or more. But when the demand is 15 cups, the conditional profit will be Rs. $[8 \times 15 - 5 \times 16 + 2 \times (16 - 15)]$ = Rs. 42. The conditional profits for each stock action-event combination can , therefore, be calculated.

Table 5.23 : Conditional Profit Table

Possible demand (event)	*Probability*	*Possible stock action (alternative)*			
	(i)	*15* *(ii)* Rs.	*16* *(iii)* Rs.	*17* *(iv)* Rs.	*18* *(v)* Rs.
15	0.10	45	42	39	36
16	0.20	45	48	45	42
17	0.40	45	48	51	48
18	0.30	45	48	51	54

The expected payoffs and the EVM for each stock action can now be calculated.

Table 5.24 : Expected Profit Table

Possible demand (event)	*Probability*	*Possible stock action (alternative)*			
	(i)	*15* *(i)×(ii)* Rs.	*16* *(i)×(iii)* Rs.	*17* *(i)×(iv)* Rs.	*18* *(i)×(v)* Rs.
15	0.10	4.50	4.20	3.90	3.60
16	0.20	9.00	9.60	9.00	8.40
17	0.40	18.00	19.20	20.40	19.20
18	0.30	13.50	14.40	15.30	16.20
EMV (Rs.)		45.00	47.40	48.60	47.40

From Table 5.24, max. EMV = Rs. 48.60 for stock action of 17 ice-cream cups each day.

Example 26:

A milkman buys milk at Rs. 2 per litre and sells for Rs. 2.50 per litre. Unsold milk has to be thrown away. The daily demand has the following probability distribution:

Demand (litres) :	*46*	*48*	*50*	*52*	*54*	*56*	*58*	*60*	*62*	*64*
Probability :	*0.01*	*0.03*	*0.06*	*0.10*	*0.20*	*0.25*	*0.15*	*0.10*	*0.05*	*0.05*

If each day's demand is independent of previous day's demand, how many litres should be ordered every day?

Solution:

Here IP = Rs. (2.50 – 2.00) = Rs. 0.50,

and IL = Rs. 200.

The milkman should stock additional litres of milk so long as the probability of selling at least an additional litre of milk is greater than p, where

$$p = \frac{IL}{IP + IL} = \frac{2.00}{2.00 + 0.50} = 0.8$$

The value of 0.8 for p implies that in order to justify the stocking of an additional unit, there must be at least 0.8 cumulative probability of selling that unit. The cumulative probabilities of sales are computed in table 9.22. Additional units should be stocked so long as the probability of selling at least an additional unit is greater than p. The optimum number of litres of milk to be stocked is 54. If the number is increased to 56, the cumulative probability will become 0.60, which is less than the required value of 0.8.

Table 5.25 : Cumulative Probability of Sales

Sales (Litres of Milk)	*Probability of this Sales Level*	*Cumulative Probability that Sales will be at this Level or Higher*
46	0.01	1.00
48	0.03	0.99
50	0.06	0.96
52	0.10	0.90
54	0.20	0.80
56	0.25	0.60
58	0.15	0.35
60	0.10	0.20
62	0.05	0.10
64	0.05	0.05

For p = 0.8,

expected incremental profit = p(IP) = Rs. 0.8 × C.50 = Rs. 0.40,

expected incremental loss = (I – P) (IL) = Rs. 0.2 × 2.00 = Rs. 0.40.

For 56 litres of stock level expected incremental loss will be more than expected incremental gain.

The use of incremental analaysis yields the same solution as provided by the EMV and EOL approaches. However, the computational effort required in this approach is much less.

Example 27:

Under an employment promotion programme it is proposed to allow sale of newspapers on the buses during off peak hours.The vendor can purchase the newspapers at a special concessional rate of 25 paise per copy against the selling price of 40 paise.Any unsold copies are,however, a dead loss.A vendor has estimated the following probability distribution for the number of copies demanded :

Number of copies :	*15*	*16*	*17*	*18*	*19*	*20*
Probability :	*0.04*	*0.19*	*0.33*	*0.26*	*0.11*	*0.07*

(a) *How many copies should he order so that his expected profit will be maximum?*

(b) *Compute EPPI.*

(c) *The vendor is thinking of spending on a small market survey to obtain additional information regarding the demand levels.How much should he be willing to spend on such a survey?*

Solution:

From the data given in the problem we can calculate the conditional profit values for each action-event combination. If CP denotes the conditional profit,S the quantity in stock and D the demand,then

CP = (40 – 25)

S = 15S,

when D ≥ S

40D – 25S,

when D < S.

The resulting conditional profit matrix is shown in table 5.26

Table 5.26 : Conditional Profit Table

Demand (event)	*Probability*	*Possible Stock Action (Alternative)*					
		15	*16*	*17*	*18*	*19*	*20*
	(i)	*(ii)*	*(iii)*	*(iv)*	*(v)*	*(vi)*	*(vii)*
		Rs.	*Rs.*	*Rs.*	*Rs.*	*Rs.*	*Rs.*
15	0.04	2.25	2.00	1.75	1.50	1.25	1.00
16	0.19	2.25	2.40	2.15	1.90	1.65	1.40
17	0.33	2.25	2.40	2.55	2.30	2.05	1.80
18	0.26	2.25	2.40	2.55	2.70	2.45	2.20
19	0.11	2.25	2.40	2.55	2.70	2.85	2.60
20	0.07	2.25	2.40	2.55	2.70	2.85	3.00

The expected payoff and EMV for each stock action can now be calculated.

Table 5.27 : Expected Profit Table

Demand (event)	*Probability (i)*	*Possible Stock Action (Alternative)*					
		15	*16*	*17*	*18*	*19*	*20*
		(i)×(ii)	*(i)×(iii)*	*(i)×(iv)*	*(i)×(v)*	*(i)×(vi)*	*(i)×(vii)*
		Rs.	*Rs.*	*Rs.*	*Rs.*	*Rs.*	*Rs.*
15	0.04	0.09	0.08	0.07	0.06	0.05	0.04
16	0.19	0.43	0.46	0.41	0.36	0.31	0.27
17	0.33	0.74	0.79	0.84	0.76	0.68	0.59
18	0.26	0.58	0.62	0.66	0.70	0.64	0.57
19	0.11	0.25	0.26	0.28	0.30	0.31	0.29
20	0.07	0.16	0.17	0.18	0.19	0.20	0.21
EMV(Rs.)		2.25	2.38	2.44	2.37	2.19	1.97

(a) The vendor should order 17 copies to get maximum expected daily profit of Rs.2.44.

Table 1.28

Event	*Probability*	*Payoff Under Perfect Information (Rs.)*	*Expected Payoff Under Perfect Information (Rs.)*
15	0.04	2.25	0.09
16	0.19	2.40	0.46
17	0.33	2.55	0.84
18	0.26	2.70	0.70
19	0.11	2.85	0.31
20	0.07	3.00	0.21
EPPI (Rs.)			2.61

Thus EPPI = Rs.2.61

(b) EPPI is computed as below.

(c) EVPI = EPPI – max EMV = Rs. (2.61– 2.44) = Rs. 0.17.

Thus the vendor should not spend more than Rs.0.17 for market survey.

Example 28:

A bicycle repairman has an opportunity to purchase a stock of discontinued bicycles. They were originally supposed to be sold Rs.400 each. The repairman is offered all five bicycles for Rs. 500, which makes his cost for each bicycles Rs.100. If he sells them he believes he can get Rs.250 for each bicycles, thereby making a profit of Rs.150. He has two options: either to buy all the discontinued bicycles or not to buy at all. There are six states of nature; these being the demand for 0, 1, 2, 3, 4 and 5 bicycles.

(i) Prepare the payoff as well as regret tables for the problem.

(ii) If the repairman has the option of buying any number of bicycles (0 to 5), find the average expected payoff and average expected regret for each stock action.

Solution

(i) The conditional profits are given by

$$CP = 150S, \text{ when } D \geq S,$$

$$250\,D - 100S, \text{ when } D < S,$$

Where S is the no. of bicycles in stock and D is the demand. The following conditional profit table is obtained:

Table 5.29 : Conditional Profit Table

Demand (event)	***Stock Action***	
	Buy none Rs.	***Buy all Rs.***
0	0	–500
1	0	–250
2	0	0
3	0	–250
4	0	–500
5	0	–750

The regret table can now be prepared.

Table 5.30 : Regret Table

Demand (event)	*Stock Action*	
	Buy none Rs.	*Buy all* Rs.
0	0	500
1	0	250
2	0	0
3	250	0
4	500	0
5	750	0

(ii) When the repairman has the option of buying any number (from 0 to 5 of bicycles, there are six stock actions (alternatives) and six events. The payoff as well as regret tables are shown below. Also average expected payoff and average expected regret for each stock action is calculated assuming the same probability 1/6 for each event.

Table 5.31 : Payoff Table

Demand (event)	*Stock Action*					
	0 Rs.	*1* Rs.	*2* Rs.	*3* Rs.	*4* Rs.	*5* Rs.
0	0	–100	–200	–300	–400	–500
1	0	150	50	–50	–150	–250
2	0	150	300	200	100	0
3	0	150	300	450	350	250
4	0	150	300	450	600	500
5	0	150	300	450	600	750
Average (Rs.)	0	108.33	175	200	183.33	125

Table 5.32 : Regret Table

Demand (event)	*Stock action*					
	0 Rs.	*1* Rs.	*2* Rs.	*3* Rs.	*4* Rs.	*5* Rs.
0	0	100	200	300	400	500
1	150	0	100	200	300	400
2	300	150	0	100	200	300
3	450	300	150	0	100	200
4	600	450	300	150	0	100
5	750	600	450	300	150	0
Average (Rs.)	375	266.67	200	175	191.67	250

Table 5.30 and 5.31 suggest an optimum stock level of 3 discontinued bicycles. This stock action results in the highest average payoff of Rs. 200 or the lowest average regret of Rs. 175.

Example 29:

A company has recently instaled new machinery but has not yet decided on the appropriate number of a certain spare part required for repairs.

Spair parts cost £ 2,000 each but are only available if ordered now. If the plant failed and there was no spare part available, the cost to the business of mending the plant rises to £ 15,000. The plant has an estimated life of 10 years and the probability distribution of failures during this time, based on the experience with similar plants, is as follows:

Table 5.33

No. of failures over ten years period	:	*0*	*1*	*2*	*3*	*4*	*5 and over*
Probability	:	*0.1*	*0.4*	*0.3*	*0.1*	*0.1*	*nil*

Calculate

(a) The expected number of failures in the ten-year period.

(b) The optimal number of spares that should be purchased now.

(c) The cost of the ordering policy chosen.

(d) The value of perfect information of the number of failures in ten-years.

Solution

(a) Expected number of failures

$= 0 \times 0.1 + 1 \times 0.4 + 2 \times 0.3 + 3 \times 0.1 + 4 \times 0.1 = 17$

(b) The cost table for each combination of number of spare parts purchased and number of failures is shown below. As an example, the combinational cost for 3 spare parts purchased and 4 failures is

$= 3 \times £\ 2{,}000 + (4 - 3) \times £\ 15{,}000$

$= £\ 21{,}000.$

Example 30:

The sales manager of Beta Co. is highly experienced in the fad market. He is sure that the sals of JUMBO (during the period it has especial appeal) will not be less than 25,000 units. Plant capacity limits total production to

a maximum of 80,000 units during JUMBO' S brief life. According to the sales manager, there are 2 chances in 5 for a sales volume of 50,000 units. The probability that it will be more than 50,000 units, is 4 times the probability that it will be less than 50,000. if sales exceed 50,000 units, volumes of 60,000 and 80,000 are equally likely. A 70,000 unit volume is 4 times as likely as either. It costs Rs. 30 to produce a unit of JUMBO whereas its selling price is estimated at Rs.50 per unit. Initial investment is estimated at Rs. 8,00,000. Should the venture of production be undertaken?

Solution

Prob. (50,000) = 2/5 = 0.40.

Prob. (less than or more than 50,000) = 0.60

Prob. (less than 50, 000) : Prob. (more than 50,000) : : 1 : 4.

Prob. (less than 50,000) = 0.60 × 1/5 = 0.12,

Prob. (more than 50, 000) = 0.60 × 4/5 = 0.48.

Prob. (60,000) : Prob. (70,000) : Prob. (80,000) : : 1 : 4 : 1.

Prob. (60,000) = 0.48 × 1/1 + 4 + 1 = 0.08,

Prob (70,000) = 0.48 × 4/1 + 4 + 1 = 0.32,

Prob. (80,000) = 0.08.

Thus we have complete probability distribution of sales except that for sales less than 50,000 we have a summarized probability of 0.12.

Table 5.34 : Payoff Table In Thousands Of Rupees

Demand (Event)	***Probability***	***Stock action*** ***<50,000 (take it 25,000) Rs.***	***50,000 Rs.***	***60,000 Rs.***	***70,000 Rs.***	***80,000 Rs.***
<50,000 (worst 25,000)	0.12	500	–250	–550	–850	–1,150
50,000	0.40	500	1,000	700	400	100
60,000	0.08	500	1,000	1,200	900	600
70,000	0.32	500	1,000	1,200	1,400	1,100
80,000	0.08	500	1,000	1,200	1,400	1,600
Expected payoff	500	850	790	690	430	

Thus the optimum expected payoff is Rs. 8,50,000. Since it is more than the initial investment of Rs. 8,00,000, the venture should be undertaken.

Example 31:

Your company manufactures goods for a market in which the technology of the products is changing rapidly. The research and development department has produced a new product which appears to have potential for commercial exploitation. A further Rs. 60,000 is required for development testing.

The company has 100 customers and each customer might purchase, at the most, one unit of the product. Market research suggests a selling price of Rs. 6,000 for each unit with total variable costs of manufacture and selling estimated at Rs. 2,000 for each unit.

As a result of previous experience of this type of market, it has been possible to derive a probability distribution relating to the proportion of customers who will buy the product, as follows:

Table 5.35

Proportion Of Customers	***Probability***
0.04	*0.1*
0.08	*0.1*
0.12	*0.2*
0.16	*0.4*
0.20	*0.2*

Determine the expected opportunity losses, given no further information than that stated above and state, whether or not, the company should develop the product.

Solution

If p denotes the proportion of customers who purchase the new product, then the conditional profit will be given by

$$CP = \text{Rs. } (6{,}000 - 2{,}000)\ (p \times 100) - 60{,}000$$

$$= \text{Rs. } 1{,}000\ (400p - 60).$$

The conditional profit table can be constructed. The company has two alternative courses of action : to develop the product or not to develop the product. From the conditional profit table, the opportunity loss table can be derived and expected opportunity losses can be found.

Table 5.36 : Conditional Profit Table

State of nature (Proportion of customers)	*Probability*	*Alternative actions*	
		Do not develop (A_1) Rs.	*Develop (A_2) Rs.*
0.04	0.1	0	– 44,000
0.08	0.1	0	– 28,000
0.12	0.2	0	– 12,000
0.16	0.4	0	4,000
0.20	0.2	0	20,000

Table 5.37 : Opportunity Loss Table

Proportion of customers	*Probability*	*Alternative actions*	
		Do not develop(A_1) Rs.	*Develop (A_2) Rs.*
0.04	0.1	0	44,000
0.08	0.1	0	28,000
0.12	0.2	0	12,000
0.16	0.4	4,000	0
0.20	0.2	20,000	0

EOL (A_1) = 0.4 (4,000) + 0.2 (20,000) = Rs.5,600.

EOL (A_2) = 0.1 (44,000) + 0.1 (28,000) + 0.2 (12,000) = Rs.9,600.

Since A_1 gives the lower EOL of Rs. 5,600, the best decision is not to develop the product.

Example 32:

A TV dealer finds that the cost of a TV in stock for a week is Rs. 30 and the cost of a unit shortage is Rs. 70. For one particular model of TV the probability distribution of weekly sales is as follows:

Weekly Sales	:	*0*	*1*	*2*	*3*	*4*	*5*	*6*
Probability	:	*0.10*	*0.10*	*0.20*	*0.25*	*0.15*	*0.15*	*0.05*

How many units per week should the dealer order ? Also find EVPI.

Solution

Here, the stocking cost = Rs. 30/ week and shortage cost = Rs.70. The cost table is constructed below. Also the expected costs for various strategies are calculated.

Table 5.38 : Cost Table

Weekly Sales	*Probability*	*Weekly stock action*						
		0 *Rs.*	*1* *Rs.*	*2* *Rs.*	*3* *Rs.*	*4* *Rs.*	*5* *Rs.*	*6* *Rs.*
0	0.10	0	30	60	90	120	150	180
1	0.10	70	30	60	90	120	150	180
2	0.20	140	100	60	90	120	150	180
3	0.25	210	170	130	90	120	150	180
4	0.15	280	240	200	160	120	150	180
5	0.15	350	310	270	230	190	150	180
6	0.05	420	380	340	300	260	220	180
Expected Cost (Rs.)		203	170	144	132	137.50	153.50	180

Thus the TV dealer should order 3 units/week. ECPI can now be prepared from cost table.

Table 5.39 : Expected Cost For Perfect Information

Demand	*Probability*	*Minimum Cost Rs.*	*ECPI*
0	0.10	0	0
1	0.10	30	3
2	0.20	60	12
3	0.25	90	22.50
4	0.15	120	18
5	0.15	150	22.50
6	0.05	180	9
			87.00

Thus EVPI = Min. expect cost – ECPI

= Rs. (132 – 87) = Rs. 45.

EXERCISES

1. What is a scientific decision-making process? Discuss the role of the statistical method in such a process.
2. Give an example of a good decision you made that resulted in a bad outcome. Also give an example of a good decision you made that had a good outcome. Why was each decision good or bad?

3. The following matrix gives the payoff (in Rs.) of different strategies (alternatives) S_1, S_2 and S_3 against conditions (events) N_1, N_2, N_3 and N_4.

	Slate of Nature			
Strategy	N_1	N_2	N_3	N_4
S_1	4,000	–100	6,000	18,000
S_2	20,000	5,000	400	0
S_3	20,000	15,000	–2,000	1,000

Indicate the decision taken under the following approaches: (i) Pessimistic, (ii) Optimistic, (iii) Equal probability, (iv) Regret, (v) Hurwicz criterion, the degree of optimism being 0.7.

4. In the Toy manufacturing company, suppose the product acceptance probabilities are not known, but the following data is known.

	Anticipated First Year Profit (Rs. 1,000s)		
Product	***Product Line***		
Acceptance	***Full***	***Partial***	***Minimal***
Good	8	70	50
Fair	50	45	40
Poor	–25	–10	0

Determine the optimal decision under each of the following decision criteria and show how you arrived at it: (a) Maximax, (b) Maximin, (c) Equal likelihood and (d) Minimax regret?

5. The following is a payoff (in rupees) table for three strategies and two states of nature:

Strategy	***State of Nature***	
	N_1	*N2*
S_1	40	60
S_2	10	–20
S_3	–40	150

Select a strategy using each of the following decision criteria: (a) Maximax, (b) Minimax regret, (c) Maximin, (d) Minimum risk, assuming equiprobable states.

6. A manufacturer's representative has been offered a new product line. If he accepts the new line he can handle it in one of the two ways. The best way according to the manufacturer would be to set a separate sales force to handle the new line exclusively. This would involve an initial investment of Rs. 1,00,000 in the office, equipment and the

hiring and training of the salesmen. On the other hand, if the new line could be handled by the existing sales force using the existing facilities, the initial investment would only be Rs. 30,000 principally for training of his present salesmen.

The new product sells for Rs. 250. The representative normally receives 20 per cent of the sales price on each unit sold of which 10 per cent is paid as commission to handle the new product. Manufacturer offers to pay 60 per cent of the sale price of each unit sold to the representative if the representative sets up a separate sales organization. Otherwise the normal 20 per cent will be paid. In either case the salesman gets a 10 per cent commission. Based on the size of the territory and experience with other products, the representative estimates the following probabilities for annual sales of the new product:

Sales (in units)	***Probability***
1,000	0.10
2,000	0.15
3,000	0.40
4,000	0.30
5,000	0.05

(a) Set up a regret table.

(b) Find the expected regret of each course of action.

(c) Which course of action would have been best under the maximin criterion?

7. Given the complete set of outcomes in a certain situation, how is the EMV determined for a specific course of action? Explain in your own words.

8. Explain the difference between expected opportunity loss and expected value of perfect information.

9. The local football club wants your advice on the number of programmes that should be printed for each game. The cost of printing and production of programmes for each game, as quoted by the local printer, is Rs. 1,000 plus 4 paise per copy. Advertising revenue which has been agreed for the season represents Rs. 800 for each game. Programmes are sold for 15 paise each. A review of sales during the previous seasons indicates that the following pattern is expected to be repeated during the coming season of 50 games:

Number of Programmes Sold	*Number of Games*
10,000	5
20,000	20
30,000	15
40,000	10

Programmes not sold at the game are sold as waste paper to a paper manufacturer at one paisa per copy.

Assuming that the four options listed are the only possibilities

(i) prepare a payoff table,

(ii) determine the number of programmes that would provide the largest profit, if a constant number of programmes were to be printed for each game,

(iii) calculate the profit which would arise from a perfect forecast of the number of programmes which would be sold at each game.

10. Indicate the difference between decision-making under risk, and uncertainty in statistical decision theory.

11. Mr. Sethi has Rs. 10,000 to invest in one of three options: A, B or C. The return on his investment depends on whether the economy experiences inflation, recession, or no change at all. His possible returns under each economic condition are given below:

		State of Nature	
Strategy	***Inflation***	***Recession***	***No Change***
A	2,000	1,200	1,500
B	3,000	800	1,000
C	2,500	1,000	1,800

What should he decide using the pessimistic criterion, optimistic criterion, equally likely criterion and regret criterion?

12. Briefly explain 'expected value of perfect information' with examples.

13. Describe a business situation where a decision-maker faces a decision under uncertainty and where a decision based on maximizing the expected monetary value cannot be made. How do you think the decision-maker should make the required decision?

14. You are given the following payoffs of three acts A_1, A_2 and A_3 and the events E_1, E_2, E_3.

States of Nature	*Three Acts*		
	A_1	A_2	A_3
E_1	25	–10	–125
E_2	400	440	400
E_3	650	740	750

The probabilities of the states of nature are respectively 0.1, 0.7 and 0.2. Calculate and tabulate EMV and conclude which of the course of action can be chosen as the best.

15. A management is faced with the problem of choosing one of three products for manufacturing. The potential demand for each product may turn out to be good, moderate or poor. The probabilities for each of the states of nature were estimated as follows:

Product	*Nature of Demand*		
	Good	*Moderate*	*Poor*
X	0.70	0.20	0.10
Y	0.50	0.30	0.20
Z	0.40	0.50	0.10

The estimated profit or loss in rupees under the three states may be taken as:

Product	*Good*	*Moderate*	*Poor*
X	30,000	20,000	10,000
Y	60.000	30,000	20,000
Z	40,000	10,000	–15,000

Prepare the expected value table, and advise the management about the choice of product.

16. The marketing staff of a certain industrial organization has submitted the following payoff table, giving profits in million rupees, concerning a certain proposal depending upon the rate of technology advance.

Technological Advance	*Decision*	
	Accept	*Reject*
Much	2	3
Little	5	2
None	(-)1	4

The probabilities are 0.2, 0.5 and 0.3 for Much, Little and None technological advance respectively. What decision should be taken?

17. A physician purchases a particular vaccine on Monday each week. The vaccine must be used within the following week, otherwise it becomes worthless. The vaccine costs Rs. 2 per dose and the physician charges Rs. 4 per dose. In the past 50 weeks, the physician has administered the vaccine in the following quantities:

Doses per week	:	20	25	40	60
Number of weeks	:	5	15	25	5

Determine how many doses the physician should buy every week.

18. Discuss the difference between decision-making under certainty, uncertainty and risk.

19. Explain the various quantitative methods which are useful for decision-making under uncertainty.

20. A grocery with a bakery department is faced with the problem of how many cakes to buy in order to meet the day's demand. The grocer prefers not to sell day-old goods in competition with fresh products; leftover cakes are, therefore, a complete loss. On the other hand, if a customer desires a cake and all of them have been sold, the disappointed customer will buy from elsewhere and the sales will be lost. The grocer has, therefore, collected information on the past sales on a selected 100-day period as shown in table below:

Sales per Day	*No. of Days*	*Probability*
25	10	0.10
26	30	0.30
27	50	0.50
28	10	0.10

Construct the payoff table and the opportunity loss table. What is the optimal number of cakes that should be bought each day? Also find and interpret EVPI (Expected Value of Perfect Information). A cake costs Re 0.80 and sells for Re 1.

21. What techniques are used to solve decision-making problems under uncertainty? Which technique results in an optimistic decision? Which technique results in a pessimistic decision?

22. A firm makes pastries which it sells at Rs. 8 per dozen in special boxes containing one dozen each. The direct cost of pastries for the firm is Rs. 4.50 per dozen. At the end of the week the stale pastries are sold off for a lower price of Rs. 3.50 per dozen. The overhead expenses

attributable to pastry production are Rs. 1.25 per dozen. Fresh pastries are sold in special boxes which cost 50 paise each and the stale pastries are sold wrapped in ordinary paper. The probability distribution of demand per week is as under:

Demand (in dozen)	:	0	1	2	3	4	5
Probability	:	0.01	0.14	0.2	0.5	0	0.05

Find the optimal production level of pastries per week.

23. The probability distribution of monthly sales of an item is as follows:

Monthly sales (units)	:	0	1	2	3	4	5	6
Probabilities	:	0.01	0.06	0.25	0.30	0.22	0.10	0.06

The cost of carrying inventory (unsold during the month) is Rs. 30 per unit per month and cost of unit shortage is Rs. 70. Determine optimum stock to minimize expected cost.

24. What is a game in game theory? What are the properties of a game? Explain the 'best strategy' on the basis of minimax criterion of optimality.
25. Explain the two-person zero-sum game giving a suitable example.
26. Explain the difference between pure strategy and mixed strategy.
27. Which competitive situation is called a game? What is the maximin criterion of optimality?
28. A producer of boats has estimated the following distribution of demand for a particular kind of boat:

No. demanded	:	0	1	2	3	4	5	6
Probability	:	0.14	0.27	0.27	0.18	0.09	0.04	0.01

Each boat costs him Rs. 7,000 and he sells them for Rs. 10,000 each. Boats left unsold at the end of the season must be disposed of for Rs. 6,000 each. How many boats should be in stock so as to maximize his expected profit?.

29. A small industry finds from the past data that the cost of making an item is Rs. 25, the selling price is Rs. 30 if it is sold within a week, and it could be disposed of at Rs. 20 per item at the end of the week:

Weekly sales	:	<3	4	5	6	7	≥ 8
No. of weeks	:	0	10	20	40	30	0

Find the optimum number of items per week the industry should produce.

30. A modern home appliances dealer finds that the cost of holding a mini cooking range in stock for a month is Rs. 200 (insurance, minor deterioration, interest on borrowed capital, etc.). Customers who cannot obtain a cooking range immediately tend to go to other dealers and he estimates that for every customer who cannot get immediate delivery, he loses an average of Rs. 500. The probabilities of a demand of 0, 1, 2, 3, 4, 5 mini cooking ranges in a month are 0.05, 0.10, 0.20, 0.30, 0.20, and 0.15 respectively. Determine the optimal stock level of cooking ranges. Also find EVPI.

31. A TV dealer finds that the cost of holding a TV in stock for a week is Rs. 50. Customers who cannot obtain new TVs immediately tend to go to other dealers and he estimates that for every customer who cannot get immediate delivery he loses an average of Rs. 200. For one particular model of TV the probabilities of a demand of 0, 1, 2, 3, 4 and 5 TVs in a week are 0.05, 0.10, 0.20, 0.30, 0.20 and 0.15, respectively.

    (i) How many televisions per week should the dealer order? Assume there is no time flag between ordering and delivery.

    (ii) Compute EVPI.

    (iii) The dealer is thinking of spending on a small market survey to obtain additional information regarding the demand levels. How much should he be willing to spend on such a survey?

32. XYZ Co. Ltd. wants to go in for a public share issue of Rs. 10 lakh (1 lakh shares of Rs. 10 each) as a part of its efforts to raise capital needed for its expansion programme. The company is optimistic that if the issue were made now it would be fully taken up at a price of Rs. 30 per share. However, the company is facing two crucial situations both of which may influence the share prices in the near future. These are

    (i) An impending wage dispute with assembly workers which could lead to a strike in the whole factory and could have an adverse effect on the share price.

    (ii) The possibility of a substantial business in the export market which would increase the share price.

    The four possible events and their expected effect on the company's share prices are envisaged as:

    E_1 : No strike, and no export business obtained—share price stays at Rs. 34.

E_2 : Strike, and ᵨort business obtained—share price stays at Rs. 30.

E_3 : No strike, and export business lost—share price hovers around Rs. 32.

E_4 : Strike, and export business lost—share price drops to Rs. 16.

And the management has identified three possible strategies that the company could adopt; viz.,

S_1 : Issue 1,00,000 shares now.

S_2 : Issue 1,00,000 shares only after the outcome of (i) and (ii) above are known.

S_3 : Issue 50,000 shares now and 50,000 shares after the outcome (i) and (ii) are known

You are required to:

(i) Draw up a payoff table for the company and determine the minimax regret solution. What alternative criteria might be used?

(ii) Determine the optimum policy for the company using the criterion of maximizing expected pay-off, given the estimate that the probability of a strike is 55 per cent and there is a 65 per cent chance of getting the export business, these probabilities being independent.

(iii) Determine the expected value of perfect information for the company.

33. A company manufacturing large electrical equipment is anticipating the possibility of a total or a partial copper strike in the near future. It is attempting to decide whether to stockpile a large amount of copper, at an additional cost of Rs. 50,000; a smᵣll amount, costing an additional Rs. 20,000: or to stockpile no additional copper at all. The stockpiling costs, consisting of excess storage, holding and handling costs and so forth, are over and above the actual material costs.

If there is a partial strike, the company estimates that an additional cost of Rs. 50,000 for delayed orders will be incurred if there is no stockpile at all. If a total strike occurs, the cost of delayed orders is estimated at Rs. 1,00,000 if there is only a small stockpile and Rs. 2,00,000 with no stockpile. The company estimates the probability of a total strike as 0.1 and that of a partial strike as 0.3.

(a) Develop a conditional cost table showing the cost of all outcomes and course of action combinations.

(b) Determine the preferred course of action and its cost. What is EPPI?

(c) Develop a conditional opportunity loss table.

(d) Without calculating EOL, find EVPI.

34. Player A is paid Rs. 8 if two coins turn both heads and Rs. 10 if two coins turn both tails. Player B is paid Rs. 3 when the two coins do not match. Given the choice of being A or B, which one would you choose and what would be your strategy?

35. Players A and B, each take out one or two matches and guess how many matches the opponent has taken. If one of the players guesses correctly, then the loser has to pay him as many rupees as the sum of the number held by both players. Otherwise, the pay out is zero. Write down the payoff matrix and obtain the optimal strategies of both players.

36. Consider a modified form of 'matching coins' game problem. The matching player is paid Rs. 8, if the two coins turn both heads and Re 1 if the coins turn both tails. The non-matching player is paid Rs. 3 when the two coins do not match. Given the choice of being a matching or non-matching player, which one would you choose and what would be your strategy?

37. What are the assumptions made in the games theory ?

38. Two computer manufacturers A and B are attempting to sell computer systems to two banks 1 and 2. Company A has 4 salesmen, company B has only 3 available. The computer companies must decide upon how many salesmen to assign to sell computer to each bank. Thus company A can assign 4 salesmen to bank 1 and none to bank 2 or three to bank 1 and one to bank 2, etc.

39. Each bank will buy one computer system. The probability that a bank will buy from a particular computer company is directly related to the number of salesmen calling from that company relative to total salesmen calling. Thus, if company A assigned three salesmen to bank 1 and company B assigns two salesmen, the odds would be three out of five that bank 1 would purchase company A's computer system. (If none calls from either company the odds are one-half for buying either computer.)

    Let the payoff be the expected number of computer systems that company A sells. (2 minus this payoff is the expected number company B sells).

What strategy would company A use in allocating its salesmen? What strategy should company B use? What is the value of the game to company A. What is the meaning of the value of the game in this problem?

40. Two candidates, X and Y, are competing for the councillor's seat in a city municipal corporation, and X is attempting to increase his total votes at the expense of Y. The strategies available to each candidate involve personal contacts, newspaper insertions, speeches or television appearance, advertising. The increase in votes available to X, given various combinations of strategies, are given below. (Assume that this is a zero-sum game, *i.e.*, any gain of X is equal to the votes lost by Y). Determine the optimal strategies that should be adopted by X during his election campaign. How many votes should X gain by adopting optimal strategy?

	Candidate Y		
Candidate X	***Personal contacts***	***Newspapers***	***Television***
Personal contacts	30,000	20,000	10,000
Newspapers	60,000	50,000	25,000
Television	20,000	40,000	30,000

41. Vishal who has an amount of Rs. 1 lakh is planning to invest it among three companies: Equity shares in company A, B and C. The payoffs in terms of (i) growth in capital, and (ii) returns on the capital are known for each of the investments under each of the three economic conditions which may prevail, that is recession, growth and stability. Assuming that Vishal must make his choice among the three portfolios for a period of one year in advance, his expectations of the net earning (in Rs.. '000) of his Rs. 1 lakh portfolio after one year is represented by the following matrix:

	Economic Condition		
Company	***Recession***	***Stability***	***Growth***
A	15	6	10
B	4	7.5	8
C	6.5	6	15

Determine the optimal strategies for investment and the expected present return for the investor under such a policy.

42. Two firms F_1 and F_2 make colour and black & white television sets. F_1 can make either 300 colour sets in a month or an equal number

of black & white sets, and make a profit of Rs. 200 per colour set and Rs. 150 per black & white set. F_2 can, on the other hand, make either 600 colour sets or 300 colour and 300 black & white sets or 600 black & white sets per month. It also has the same profit margin on the two sets as F_1. Each month there is a market of 300 colour sets and 600 black & white sets and the manufacturers would share market in the proportion in which they manufacture a particular type of sets.

Write the payoff matrix of F_1 and F_2 per month. Obtain F_1 and F_2's optimal strategies and the value of the game.

43. In zero-sum two-person children's game of stone, paper and scissors, both players simultaneously call out stone paper or scissors. The combination of paper and stone is a win of one unit for player calling paper (paper cover; stone); stone and scissors is a win for stone (stone breaks scissors), and scissors and paper is a win for scissor; (scissors cut paper). A call of the same item represents no pay-off. Write the pay-off matrix and the equivalent linear programming problem to the above game. Find the optimal strategy for both the players and the value of the game.

44. Obtain the strategies for both players and the value of the game for two-person zero-sum game whose payoff matrix is given as follows:

(a)

Player A	***Player B*** B_1	B_2	B_3
A_1	1	3	11
A_2	8	5	2

(b)

Player A	***Player B*** B_1	B_2	B_3	B_4	B_5	B_6
A_1	1	3	–1	4	2	–5
A_2	–3	5	6	1	2	0

45. Obtain the optimal strategies for both players and the value of the game for two-person aero-sum game whose payoff matrix is given as follows:

Player A	***Player B*** B_1	B_2
A_1	–6	7
A_2	4	–5
A_3	–1	–2
A_4	–2	5
A_5	7	–6

46. For the following payoff tables, transform the zero-sum game into an equivalent linear programming problem and solve it by simplex method.

(a)

	Player B		
Player A	B_1	B_2	B_3
A_1	9	1	4
A_2	0	6	3
A_3	5	2	8

(b)

	Company A		
Company B	A_1	A_2	A_3
B_1	2	–2	3
B_2	–3	6	–1

47. A soft drink company calculated the market share of two products against its major competitor having three products and found out the impact of additional advertisement in any one of its products against the other.

	Company B		
Company A	B_1	*A*	B_3
A_1	6	7	15
A_2	20	12	10

What is the best strategy for the company as well as the competitor? What is the payoff obtained by the company and the competitor in the long run? Use graphical method to obtain the solution.

48. Two Firms A and B make colour and black & white television sets. Firm A can make either 150 colour sets in a week or an equal number of black and white sets, and make a profit of Rs. 400 per colour set and Rs. 300 per black & white set. Firm B can, on the other hand, make either 300 colour sets, or 150 colour and 150 black and white sets, or 300 black & white sets per week. It also has the same profit margin on the two sets as A. Each week there is a market of 150 colour sets and 300 black & white sets and the manufacturers would share market in the proportion in which they manufacture a particular type of set.

Write the payoff matrix of A per week. Obtain graphically A's and B's optimal strategies and value of the game

49. A car manufacturer uses a special control device in each ear he produces. Two alternative methods can be used to detect and avoid a faulty device. Under the first method, each device is tested before it is installed. The cost is Rs. 2 per test. Alternatively, the control device can be installed without being tested, and a faulty device can be detected and rendered after the car has been assembled, at a cost of Rs. 20 per faulty device.

Regardless of which method is used, faulty devices cannot be repaired and must be discarded. A manufacturer purchases the control devices in batches of 10,000. Based on past experience, he estimates the proportion of defective components and the associated probability to be:

Proportion of Faulty Devices	*Probability*
0.08	0.20
0.12	0.70
0.16	0.10

(a) Which inspection method should the manufacturer adopt?

(b) What is the expected value of perfect information (EVPI)?

50. Raman Industries Ltd. has a new product which they expect has great potential. At the moment they have two courses of action open to them: To test market (S_1) and to drop product (S_2). If they test it, it will cost Rs. 50,000 and the response could be positive or negative with probabilities 0.70 and 0.30, respectively. If it is positive, they could either market it with full effort or drop the product. If they market with full scale, then the result might be low, medium, or high demand and the respective net payoffs would be – Rs. 1,00,000, Rs. 1,00,000 or Rs. 5,00,000. These outcomes have probabilities of 0.25,0.55 and 0.20, respectively.

    If the result of the test marketing is negative they have decided to drop the product. If, at any point, they drop the product there is a net gain of Rs. 25,000 from the sale of scrap. All financial values have been discounted to the present. Draw a decision tree for the problem and indicate the most preferred decision.

51. A manufacturing company has just developed a new product. On the basis of past experience, a product such as this will either be successful, with an expected gross return of Rs. 1,00,000, or unsuccessful, with an expected gross return of Rs. 20,000. Similar products manufactured by the company have a record of being successful about 50 per cent of the time. The production and marketing costs of the new product are expected to be Rs. 50,000.

    The company is considering whether to market this new product or to drop it. Before making its decision, a test marketing effort can be conducted at a cost of Rs. 10,000. Based on past experience, test marketing results have been favourable about 70 per cent of the time. Furthermore, products favourably tested have been successful 80 per cent of the time. However, when the test marketing result has been unfavourable, the product has only been successful 30 per cent of the time. What course of action should the company pursue?

52. A well-known departmental store advertises female fashion garments from time to time in the Sunday press. Only one garment is advertised on each occasion. Experience of garments in the price range Rs. 30-40 leads the management to assess the statistical probability of demand at various levels after each advertisement as follows:

Demand (No. of garments)	*After Advertising in One Newspaper*	*After Advertising in Two Newspapers*
30	0.10	0.00
40	0.25	0.15
50	0.40	0.35
60	0.25	0.40
70	0.00	0.10

The next garment to be advertised at Rs. 35 will cost Rs. 15 to make and can be disposed of to the trade, if unsold, for Rs. 10. The store's advertising agents will charge Rs. 37.50 for artwork and blockmaking for the advertisement and each newspaper will charge Rs. 50 for the insertion of the advertisement. You are required to:

(a) Calculate how many garments to the nearest 10 should be purchased by the store in order to maximize expected gross profit, after advertising in:

(i) one newspaper,

(ii) two newspapers.

(b) Calculate the amount, if any, of the expected net profit after advertising:

(i) in one newspaper, (ii) two newspapers.

53. The XYZ Company manufacture guaranteed tennis balls. At present time, approximately 10 per cent of the tennis balls are defective. A defective ball leaving the factory costs the company Re 0.50 to honour its guarantee. Assume that all defective balls are returned. At a cost of Re 0.10 per ball, the company can conduct a test, which always correctly identifies both good and bad tennis balls.

(a) Draw a decision tree and determine the optimal course of action and its expected cost.

(b) At what test cost the company should be indifferent to testing?

54. The Ore Mining Company is attempting to decide whether or not a certain piece of land should be purchased. The land cost is Rs. 3,00,000. If there are commercial ore deposits on the land, the estimated value of property is Rs. 5,00,000. If no ore deposits exist, however, the property value is estimated at Rs. 2,00,000. Before purchasing the land, the property can be cored at a cost of Rs. 20,000. The coring will indicate if conditions are favourable or unfavourable for ore mining. If the coring report is favourable the probability of recoverable ore deposits on the land is 0.8, while if the coring report is unfavourable the probability is only 0.2. Prior to obtaining any coring information, management estimates that the odds are 50-50 that ore is present on the land. Management has also received coring reports on pieces of land similar to the ore in question and found that 60 per cent of the coring reports were favourable.

Construct a decision tree and determine whether the company should purchase the land, decline to purchase it, or take a coring test before making its decision. Specify the optimal course of action and EMV.

55. Suppose a company has several independent investment opportunities each of which has an equal chance of gaining Rs. 1,00,000 or losing Rs. 60,000. What is the probability that the company will lose money on two such investments? On three such investments? On four such investments?

If a company has a number of independent investment opportunities, in each of which the financial risk is relatively small compared to its overall asset position, why should the company try to maximize EMV, rather than expected utility?

56. An oil drilling company is considering the purchase of mineral rights on a property of Rs. 100 lakhs. The price includes tests to indicate whether the property has type A geological formation or type B geological formation. The company will be unable to tell the type of geological formation until the purchase is made. It is known, however, that 40 per cent of the land in this area has type A formation and 60 per cent type B formation. If the company decides to drill on the land it will cost Rs. 200 lakh. If the company does drill it may hit an oil well, gas well or a dry hole. Drilling experience indicates that the probability of striking an oil well is 0.4 on type A and 0.1 on type B formation. Probability of hitting gas is 0.2 on type A and 0.3 on type B formation. The estimated discounted cash value from an oil well is Rs. 1,000 lakh and from a gas well is Rs. 500 lakh. This

includes everything except cost of mineral rights and cost of drilling. Use the decision-tree approach and recommend whether the company should purchase the mineral rights?

57. XYZ company dealing with a newly invented telephone device is faced with the problem of selecting out of the following courses of action available:

    (i) manufacture the device itself; or

    (ii) be paid on a royalty basis by another manufacturer; or

    (iii) sell the rights for its invention for a lump sum.

    The profit (in Rs'000s) which can be expected in each case and the probabilities associated with the level of sales are shown in the following table:

Outcome	*Probability*	*Manufacture Itself*	*Royalties*	*Sell All Rights*
High sales	0.1	75	35	15
Medium sales	0.3	25	20	15
Low sales	0.6	10	10	15

Represent the company's problem in the form of a .decision tree. Redraw further the decision tree by introducing the following additional information:

(a) If it manufactures itself and sales are medium or high, then company has the opportunity of developing a new version of its telephone;

(b) From past experience company estimates that there is a 50 per cent chance of successful development,

(c) The cost of development is Rs. 15 and returns after deduction of development cost are Rs. 30 and Rs. 10 for high and medium sales, respectively.

58. A company has developed a new product in its R&D laboratory. The company has the option of setting up production facility to market this product straight-away. If the product is successful, then over the three years expected product life, the returns will be Rs. 120 lakh with a probability of 0.70. If the market does not respond favourably, then the returns will be only Rs. 15 lakh with probability of 0.30.

    The company is considering whether it should test market this product building a small pilot plant. The chance that the test market will yield favourable response is 0.80. If the test market gives favourable response, then the chance of successful total market improves to 0.85.

If the test market gives poor response, then the chance of success in the total market is only 0.30.

As before, the returns from a successful market will be Rs. 120 lakh and from an unsuccessful market only Rs. 15 lakh. The installation cost to production is Rs. 40 lakh and the cost of test marketing pilot plant is Rs. 5 lakh. Using decision-tree analysis, draw a decision tree diagram, carry out necessary analysis to determine the optimal decisions.

59. The Sensual Cosmetic Co. has developed a new perfume which management feels has a tremendous potential. It not only interacts with the wearer's body chemistry to create a unique fragrance but is also especially long lasting. A total of Rs. 10 lakhs has already been spent on its development. Two marketing plans have been devised:

(a) The first plan follows the company's usual policy of giving small samples of the new products when other items in the company's product lines are purchased and placing advertisements in women's magazines. The plan would cost Rs. 5 lakhs and it is believed that it might result in a high, moderate or low market response with probability of 0.2, 0.5 and 0.3, respectively. The net profit excluding development and promotion costs in these cases would be Rs. 20 lakh, Rs. 10 lakhs and Re 1 lakh, respectively. If it later appears that the market response is going to be low it would be possible to launch a TV advt. campaign. This would cost another Rs. 7.5 lakh. It would change the market response to high or moderate as previously described but with probability of 0.5 each.

(b) The second marketing plan is much more aggressive than the first. The emphasis would be heavily upon TV advertising. The total cost of this plan would be Rs. 15 lakhs, but the market response would be either excellent or good, with probabilities of 0.4 and 0.6, respectively. The profit excluding development and promotion costs would be Rs. 30 lakhs and Rs. 25 lakhs for the two outcomes.

Advise on the sequence of strategy to be followed by the company.

60. The investment staff of TNC Bank is considering four investment proposals for a client: shares, bonds, real estate and saving certificates. These investments will be held for one year. The past data regarding the four proposals are given below:

Shares : There is a 25 per cent chance that shares will decline by 10 per cent, a 30 per cent chance that they will remain stable and a 45 per cent chance that they will increase in value by 15 per cent. Also the shares under consideration do not pay any dividends.

Bonds : These bonds stand a 40 per cent chance of increase in value by 5 per cent and 60 per cent chance of remaining stable and yield 12 per cent.

Real Estate : This proposal has a 20 per cent chance of increasing 30 per cent in value, a 25 per cent chance of increasing 20 per cent in value, a 40 per cent chance of increasing 10 per cent in value, a 10 per cent chance of remaining stable and a 5 per cent Chance of losing 5 per cent of its value.

Savings Certificates These certificates yield 8.5 per cent with certainty.

Use a decision tree to structure the alternatives to the investment staff, and using the expected value criterion, choose the alternative with the highest expected value.

61. The corporation has recently got leasehold drilling rights on a large area in the western pan of the country. No seismic coverage is available and to conduct a detailed survey Rs. 3 million is required. If oil is struck, a large reserve may result in a net profit of Rs. 30 million, whereas a smaller marginal reserve may result in a net profit of Rs. 18 million. Cost of drilling wildcast well is Rs. 7 million.

Seismic is thought to be quite reliable in this area. Uncertainty involved is whether or not a structure exists. The company assesses probability that the test producing good, fair or bad result is 0.40, 0.30 and 0.30, respectively. On the basis of past drilling records and experiences indicating the probabilities of striking oil in large reserve, smaller marginal reserve or dry hole even in the presence of good, fair and bad reading of seismic study are as under:

Seismic Study	***Probability of Yield***		
	Large Reserve	***Marginal Reserve***	***Dry Hole***
Good	0.50	0.25	0.25
Fair	0.30	0.30	0.40
Bad	0.10	0.20	0.70

Exploratory group has suggested two possible exploration strategies:

(i) Drill at once on the basis of present geologic intepretation and extrapolations.

(ii) Conduct a seismic study and defer drilling till seismic data is reviewed.

As a member of strategic group of the company evaluate the two strategies suggested by the exploratory group.

62. An automobile owner faces the decision as to which deductible amount of comprehensive insurance coverage to select. Comprehensive coverage includes losses due to fire, vehicle theft, vandalism and the forces of nature. The possible choices are zero deductible coverage for Rs. 60 per year or Rs. 50 deductible coverage for Rs. 45 per year. (The owner pays the first Rs. 50 of any loss of at least Rs. 500.) Considering accidents covered by the comprehensive portion of the policy some of the owner's utility values are given below:

Amount (Rs.)	:	–95	–60	–50	–45	0
Utility	:	0.2	0.4	0.45	0.47	0.5

(a) Sketch the owner's utility curve. Is the owner a risk avoider, an EMV'er or a risk taker?

(b) On the basis of the utility curve drawn in part (a), should the owner take the zero deductible or the Rs. 50 deductible comprehensive coverage?

MULTIPLE CHOICE QUESTIONS

1. A type of decision-making environment is
   (a) certainty (b) uncertainty
   (c) risk (d) all of the above
2. Decision theory is concerned with
   (a) methods of arriving at an
   (b) selecting optimal decision in sequential manner, optimal decision.
   (c) analysis of information that is available.
   (d) all of the above.
3. Games which involve more than two players are called
   (a) conflicting games (b) negotiable games
   (c) N-person games (d) all of the above
4. The decision-maker's knowledge and experience may influence the decision-making process when using the criterion of
   (a) maximax (b) minimax regret
   (c) realism (d) maximin

5. The difference between the expected profit under conditions of risk and the expected profit with perfect information is called
   (a) expected value of perfect information
   (b) expected marginal loss
   (c) expected opportunity loss
   (d) none of the above
6. When the sum of gains of one player is equal to the sum of losses to another player in a game, this situation is known as
   (a) biased game (b) zero-sum game
   (c) fair game (d) all of the above
7. Two person zero-sum game means that the
   (a) sum of losses to one player equals the sum of gains to other
   (b) sum of Jossès to one player is not equal to the sum of gains to other
   (c) both (a) and (b)
   (d) none of the above
8. A game is said to be fair if
   (a) both upper and lower values of the game are same and zero
   (b) upper and lower values of the game are not equal
   (c) upper value is more than lower value of the game
   (d) none of the above
9. The minimum expected opportunity loss (EOL) is
   (a) equal to EVPI (b) minimum regret
   (c) equal to EMV (d) both (a) and (b)
10. The expected value of perfect information (EVPI) is
    (a) equal to expected regret of the optimal decision under risk
    (b) the utility of additional information
    (c) maximum expected opportunity loss
    (d) none of the above
11. The value of the coefficient of optimism (a) is needed while using the criterion of
    (a) equally likely (b) maximin
    (c) realism (d) minimax

12. The concept of utility is used to
    (a) measure the utility of money
    (b) take into account aversion of risk
    (c) both (a) and (b)
    (d) none of the above
13. Which of the following criteria is not applicable to decision-making under risk?
    (a) maximize expected return
    (b) maximize return
    (c) minimize expect regret
    (d) knowledge of likelihood occurrence of each state of nature
14. Which of the following criteria is not used for decision-making under uncertainty?
    (a) maximin (b) maximax
    (c) minimax (d) minimize expected loss
15. Game theory models are classified by the
    (a) number of players (b) sum of all payoffs
    (c) number of strategies (d) all of the above